I0473553

PHILOSOPHIÆ NATURALIS

PRINCIPIA MATHEMATICA

Revision IV – Vol II

By

Isaac Newton

And

Keith Dixon-Roche

PHILOSOPHIÆ NATURALIS

PRINCIPIA MATHEMATICA

Revision IV – Vol II

Published by CalQlata
info@CalQlata.com

First published November 2018
Second publication February 2019
Third publication June 2019
Fourth publication December 2019
Copyright © Keith Dixon-Roche 2018

Contents

Preface

I have always believed that if a mathematical law applies to one feature of nature it must apply to all of it: i.e. a law must by definition, be universal. I also feel that science took a wrong turning in the first quarter of the twentieth century owing to the dissemination of highly speculative theories that were accepted simply because of the prominence of their proposers. However, I was not sufficiently familiar with the subject to dispute it. After two and a half years of detailed study, that situation has changed and it appears to me that a hundred years may have been wasted in the search for impossible solutions. Isaac Newton's laws should have prevailed.

Newton apparently devised his theories to settle a bet, and like everything he tackled he took this work seriously. Despite having only Kepler's elliptical orbits and Galileo's laws of motion at his disposal, Newton managed to develop an all-encompassing theory that remains universally valid today. It was published in three revisions between 1687 and posthumously. He published only because of the persistence of one of his few friends: Edmund Halley. It is for this, rather than for his comet, that we owe Edmund Halley our deepest gratitude.

Newton's gravitational theory is complete and totally accurate. It covers all the bases. His model relies on a concept he called his 'constant of motion' to keep things moving. However, even he didn't realise that his theory also applies to circular orbits in which a satellite (e.g. an electron) provides its own kinetic energy, or that his gravitational constant (G) may be used to calculate the deflection of electro-magnetic radiation (light).

His laws of gravitation and motion together describe the behaviour of everything in the universe from atomic particles to the Big Bang, and they do so with absolute simplicity and accuracy, except for one small omission; he did not explain spin theory, without which it is difficult to explain all motion. However, it would have been very difficult for him to have developed this theory with the limited information and facilities available to him.

A couple of years ago, my daughter gave me a copy of Colin Pask's *Magnificent Principia* (Pask; 2013). After reading it, I was left with the suspicion that there were many unanswered questions about Newton's discoveries and I wondered how much had been done to continue Newton's work during the subsequent 300 years. Very little it seems.

I therefore set out on my quest to prove every aspect of Newton's theory of orbital motion, and see if I could determine the source of planetary spin. Having completed these objectives, I continued with core pressure, the earth's magnetic field, the definition of his gravitational constant 'G' and finally the atom, all using his theories.

After completing my model of the atom and having discovered how it really works, I was stunned by its simplicity and brilliance. Its existence must surely be due to providence, not chance. If there is one thing that could prove the existence of a being of supreme intelligence, and I am not referring to anybody's particular god; it is the atom. In the immortal words of a great contemporary philosopher; I was that *girl sitting on her own in a small café in Rickmansworth*" (Adams; 1980), and I couldn't understand why none of this had been done before.

I am an engineer, not a scientist. Whilst I have always had an interest in science, I never had the opportunity to study it in detail. As a non-scientist, I have been able to tackle the subject free from the dogma that the scientific community has acquired since it displaced the religious community's hegemony over its own flawed natural laws.

Whilst my theories and models may not be perfect, everything in them can be supported with known scientific theories evolved well before the twentieth century.

I realise of course, that just as with Copernicus, Kepler, Galileo, Newton and Wegener before, none of these findings will be appreciated whilst the current scientific community exists. That august body is hardly likely to accept theories that disprove those for which they have been awarding themselves so many prestigious prizes. My hope is that maybe, one day, a new generation of free-thinking scientists will discover this work, correct, complete and advance it, and in so doing get science back on track.

Because it is now possible to define the Milky-Way's force-centre, I have given it the name 'Hades' for easier reference.

Keith Dixon-Roche 2018

1 Introduction

Science got itself into a bit of a mess during the 20th century owing to a couple of obscure theories, neither of which can be reconciled with concepts that we *know* work, but which stubbornly refuse to go away. Together these theories have inspired countless myths that simply multiply with the passing years. Nobody appears to be questioning them and nobody is able to verify them.

It has now become standard practice within the scientific community to justify any irreconcilable theory simply by claiming that *"the laws of physics do not apply"*.

So, I decided to have a go myself, by starting all over again; going back to basics (the year 1900).

Apart from Max Planck's assistance, I have managed to sort out this mess and compile a complete working theory for the universe using principles that were available well before 1900.

I have also managed to describe all the universal constants (including electrical) in the same basic units of energy (mass, length, time, charge & temperature; refer to Chapter 5)

Anything in this book that has not been fully resolved (and there isn't much) is referred to as *hypothesis*.

Whilst my hypotheses are perfectly robust, they remain as such because a couple of details need confirming/correcting. The contentious aspects mostly involve the nature of neutrons, but this has not been addressed here because it has nothing to do with Newton's laws of orbital motion.

Unresolved issues are highlighted in the text with the superscript [?] in which '?' will be replaced with a number that can be found in Chapter 7

1.1   What Went Wrong

Unfortunately, about a hundred years ago, a prominent scientist stated of his own theoretical model: *"if you aren't profoundly shocked by quantum physics, then you haven't understood it"*
Another did not appear to understand the basis for Henri Poincaré's formula E=mc² and actually declared to Georges Lemaître that his (Lemaître's) *"science was not very good"*

Such comments should be treated with extreme caution ...
... if there is one thing certain about nature, it does not need to rely on complexity for an elegant solution, and scientific laws *never* rely on statistics because statistics are subject to change; *laws are not.*
Statistics are akin to chaos theory: they are a means of guesswork used in situations where insufficient information is available to explain events accurately. They apply to the consequences of laws, not the laws themselves.

Quantum theory is inelegant, over complicated, reliant on statistics, cannot be reconciled with Newton's laws of orbital motion, cannot emit energy and remains unresolved after a hundred years. It is highly likely therefore, that it is nothing more than an obscure theoretical exercise.

Relativism can be disproved using Newton's gravitational constant and dark matter remains undiscovered. Poincaré's formula has nothing to do with kinetics. Classical atomic theory appears to be incorrect. Black-holes are wrongly said to be singularities that spin at the speed of light. Nobody has tried to determine the source of planetary (and therefore atomic) spin or core-pressure.

Isaac Newton pointed us in the right direction 300 years ago, but since the early 20th century the entire scientific community seems to have discounted the suitability of his theories for the evaluation of atoms (quantum theory) and galaxies (dark matter) simply because a couple of well-known scientists took this view. For example:

Relativism appears to have been partly based upon the supposition that E=mc² applies to kinematics, whereas it is a limiting case for potential energy based upon Newton's and Coulomb's laws and the creation of neutrons. Moreover, it incorrectly assumes that light possesses mass.

It is incorrectly currently believed that mass converts to energy with speed.

Dark matter in the form of sub-atomic particles was postulated because Newton's laws were said to predict a great deal more matter in the Milky Way than appears through observation. This has been easily disproved.

It was long ago assumed that we need sub-atomic particles (e.g. quarks, leptons, fermions, bosons, gluons, etc.) to hold atomic particles together and make the atom work. It now appears that none of these are necessary.

We have been taught that atomic shells are elliptically flat, can hold more than two electrons, and that each electron within a shell is in some way different from all others. It now appears that this level of complexity is unnecessary.

As Newton's gravitational constant (G) is based upon Quanta why shouldn't his theories also apply to atoms?

We have been advised by the world's most eminent astrophysicists that it is impossible to calculate spin in satellites and force-centres. Yet Newton's laws provide us with all the information needed to solve this problem.

Nobody appears to have grasped the fact that Newton's formula directly (with no reinterpretation) allows us to calculate the pressure inside a solid body, such as a planet or star, so why are we still guessing internal pressures?

In fact, guesswork appears to be prevalent throughout science today.

Together with the help of a number of early heroes (refer to Appendix A8), Newton provided everything we need to understand our universe ...
... how it was created,
... the age of everything in it,
... how it works,
... what everything in it is made of,
... how it generates its energy and
... where it stores this energy;

The end of the 18th century saw the start of the industrial revolution, which continues today, only now; it is called a technological revolution. The start of the 20th century should have kicked off a scientific revolution. It never happened. Why?

1.1.1   The Photon

The problem was the photon.

I need to deal with this issue now in order that it doesn't interfere with your understanding of the universal model discussed in this book.

It is about time we all dropped the concept of photons, i.e. the belief that electrons travelling at the speed of light emit light; *they don't*.
We have been taught this for a hundred years, forcing us to create weird and wonderful theories to explain how *mass* moves in waves; *it doesn't*.
The photon exists in our minds because of a very simple mistake made a long time ago related to Crooke's tube (refer to Chapter 6.4).

Once this is understood, the whole problem of energy, magnetism, gravity, electricity, etc. vanishes. You can ignore quantum theory and the theory of relativity, both of which were invented to explain this misunderstood behaviour of electrons.

The deflection of light can *only* be explained using Newton's gravitational constant (G), and the behaviour of electrons within atoms can *only* be resolved using Newton's laws of orbital motion and Coulomb's laws of electrical force (refer to Chapters 6.2.1 & 6.11.2). We should not, however, forget William Gilbert's contribution, which predates and forms the basis of all the theories related to force and energy fields (both atomic and astronomic).

It appears to me that if everybody had realised that Crooke could not possibly have created a perfect vacuum in his tube, we would not have been confused by quantum theory and the theories of relativity, and we would now be 100 years into a 'scientific revolution'.

1.2   And Now?

Whilst the theories proposed in this book concerning Newton's Laws of Orbital Motion, Orbital Systems, Planetary Spin, Core Pressure, the Atom and Earth's Magnetic Field are a matter of scientific fact, those on Energy and the universe are hypotheses.

However, they ...
... are based on and obey well-known universal laws of nature that work
... have no need for statistics, unification theories or obscure concepts
... reflect what we sense in the universe
... have no need for intimidation

It cannot have escaped everyone's notice that Newton's, Coulomb's, Gilbert's, Maxwell's and others' force formulas all have the same configuration:
$F = K.v_1.v_2 / R^2$ (which is actually: $F = K.v_1.v_2 / A$)
where: 'K' is a constant, 'v' a variable and 'A' the spherical surface area at radius (R).

My own calculations have revealed a similar relationship for the conversion of electro-magnetic energy to velocity in electrons:
$T = X.v^2 / e^2$
where 'X' is a constant, 'v' the velocity of an electron and 'e' its electrical charge.

If all these formulas *look* the same, they probably *are* the same, i.e. they are simply variations based upon our current misunderstanding of gravity, mass, heat, etc. which are actually the same thing; *energy*. Thus, there are really only two formulas, one of which is for electrical force (Coulomb) and the other for magnetic force (Gilbert/Newton) that differ by a coupling ratio ($\varphi = 4.407E\text{-}40$). Given that gravity is magnetism, we need be in no doubt that Newton's formula represents magnetic force and can be explained as such (refer to Chapter 6.8).

The atomic model proposed here is elegant, eternal, predictable and brilliantly simple; anyone can understand it without the need for shock-tactics. It also complies with all of Newton's, Gilbert's, Coulomb's, Faraday's and Maxwell's laws, so there is no need for unification theories or statistics. In fact, it now looks highly likely that contrary to popular scientific opinion, these laws are sufficient to explain everything in our universe. Newton's laws are indeed universal, and via them, we can create realistic solutions for virtually everything in our universe from atomic to astronomic physics, including: neutronic energy, 'Big-Bang', Earth's magnetic field, 'G', 'E=mc²', ultimate density and a great deal more.

Everything is energy: our universe is very much simpler than the one we have been taught, and exploited properly it can provide us with all the clean, free energy we need, simply from Newton's orbits.

If my model is correct (or even close), it then becomes a simple, albeit time-consuming enterprise to determine everything there is to know about our universe, from the very smallest (Quanta) to the very largest (*Big-Bang*) using theories that have been known since Poincaré first revealed his formula and Crooke discovered electro-magnetic energy in the 19th century.

1.3   Where Do We Go From Here?

Given what we now know about universal energy;
1) How it is created (orbits and spin-friction)
2) Where it is created (stars and planets)
3) How it is transmitted (electro-magnetic energy)
4) Where it is stored (neutrons)

We now have access to unlimited, clean, free energy sources;
1) Elliptical Orbits
2) Mantle heat
3) Neutrons

Moreover, these theories can give us the ability to *mathematically* predict chemical reactions in *all* matter irrespective of complexity; the *ultimate calculator*.

Such a calculator would preclude the need for material, chemical or pharmaceutical testing and experimentation. No more risk, material, time or money need be wasted on such activities and every country in the world would be able to design [100% accurate] new materials, chemicals and medicines in safety, from a computer terminal with trained but semi-skilled personnel. Furthermore, the creation of comprehensive organic and inorganic chemical databases will remove the need for duplicate effort together with the horrendous qualification periods for new medicines imposed by various national and international health authorities.

Because we now know where the universe stores its energy, we have access to an unlimited supply free from waste and pollution. We could do something useful with the world's nuclear waste; as the fuel for clean, controllable, efficient energy generators of any size. Much less mining!

Moreover, due to the discovery of the true meaning of $E=mc^2$ (refer to Chapter 6.2.5), there is no longer any reason to assume that light-speed is a limiting condition for matter. And if matter has no mass, imposing a limiting velocity owing to the conversion of mass to energy becomes unnecessary. The speed of light is simply a speed for electro-magnetic radiation, such as that for sound: there's no reason it cannot be exceeded.

Anti-gravity also becomes *theoretically* possible. All you need to do is repel the earth's *magnetism*, which is easier than opposing *gravity* with mass.

A few of the possibilities from the discoveries explained in this book are listed below?

1) Molecular calculator (and database) giving new (perfect) materials, medicines and chemicals in minutes
2) Clean, free efficient energy (by-product = hydrogen)
3) Propulsion-free satellites
4) The ability to safely recycle nuclear waste
5) Energy cells that can be fuelled with any matter (e.g. rocks!)
6) Alter elements into something else
7) Change the colour of matter electrically
8) Together with PERS#, the elimination of skin-friction offers virtually free travel
9) Perfect lubricants (machines with almost eternal life)
10) Free energy from the earth's mantle
11) Massive reductions in: pollution, material waste, energy, etc.
PERS = potential energy recovery system

In other words, we now have the ability to ...
... massively reduce energy and battery production;
... massively reduce mining requirements;
... massively reduce transport costs;
... massively reduce the number of chemical laboratories;
... eliminate; national power stations & transmission lines, wind-turbines & solar panels;
... eliminate pollution from energy generation;
... create vehicles with no engine or drivetrain that need no refuelling;
... create 100% recyclable packaging

All the energy we use today requires the generation of much more to harness and recycle it. Instead of generating energy at an efficiency of less than 10%, we now have access to energy generation that is 231,000,000% efficient.

Instead of swapping one pollutant for another and/or simply moving it around as we do today, we could now create a genuinely clean place for everyone in which to live; together with limitless cheap energy for all.

1.4   How This Book Is Organised

This volume provides a non-technical description of this universal model:

2 Narrative

A written description that gives a general overview of the various discoveries made in this book. It is devoid of formulas and mathematical complexity with a view to providing a *'light-read'*!

3 Calculation Procedures (Vol II)

A compilation of the mathematical formulas supporting the narrative, including how to use them. This section has been written to simplify their use.

4 Calculation Results (Vol II)

A collection of [mostly] tabulated calculation results for selected examples using the formulas provided in section 3 (above).

5 Physical Constants (Vol III)

All the physical constants (including electrical properties such as Volts, Amps, Henries, Farads, Ohms, etc.) are provided (to ≤15 decimal places) in terms of the same four basic units; length, time, mass and charge and two ratios: m_e, e, R_n, t_n & ξ_v, ξ_m

6 Support (Vol III)

A mathematical and descriptive explanation for all the physical constants and scientific discoveries along with the reasons why Relativity and Quantum Theory must now be discarded.

7 Things You Can Do! (Vol III)

A list of unresolved issues.

8 Appendices

References, symbols, glossary, etc. used throughout this book along with a summary list of corollaries and hypotheses.

For reference purposes, the contents List and page numbering in volumes I, II & III match the page numbering in the complete book.

3 Calculation Procedures

A compilation of the mathematical formulas supporting the narrative, including how to use them. This section has been written to simplify their use.

3.1 Energy

The calculation procedures for magnetic and electrical energy have been included in the **Support** section of this book, along with heat, gravity, mass etc. (Refer to Chapters (6.6 to 6.9)

3.1.1 Electrical

Refer to Chapter 6.9

3.1.2 Magnetic

Refer to Chapters 6.7 & 6.8

3.1.3   Electro-Magnetic

Electro-magnetic radiation always travels at the same **velocity**, which we currently refer to as the *speed of light*: $c = 299792459$ m/s, but is of course, the same speed as *all* electro-magnetic energy.

Knowing this, and that its energy is exactly the same as the kinetic energy in the electron that transferred it, we can calculate its other properties:

Wavelength (λ) is defined by the orbital velocity (v) of the electron transferring it: $\lambda = 2\pi R.c/v = c/f$ {m}
Where: R is the electron orbital (shell) radius

Johannes Rydberg gave us a relationship between wavelength and shell number (n) that actually works:
If the electron temperature (T_1) in shell number 1 (n=1) is known, the electron temperature (T_n) in any shell may be calculated as follows:
$T_n = T_1/n$
For example; the electron temperature at the Bohr [orbital] radius (a_o) which was actually discovered by Rydberg ...
$T = X_R/a_o = 33192.4000063507$ K
Rydberg's constants may be used to calculate the electro-magnetic wavelength and energy generated by an electron in any shell as follows:
$E_n = R_y . T_n/T$
$\lambda_n = \frac{1}{2} . (T/T_n)^{1.5} / R_\infty$
Where: X_R (heat transfer coefficient) $= 1.75646616508036E\text{-}06$ K.m

Frequency (f) is defined by the orbital period of the electron transferring it: $f = v/2\pi R = c/\lambda$ {Hz}

Amplitude (A) is equal to the orbital radius of the electron transferring it: $A = R$ {m}

Energy (E) may be determined using the modified version of Planck's constant or the orbital velocity (v) of the electron transferring it:
$E = h'/A = \frac{1}{2}.m_e.v_e^2$ {J}

A proton develops an operational **Charge** (e') whilst hosting an orbiting electron that maximises at: $e_m = e.\xi_v$
$e' = e . v/v_o$ {C}
This charge is used to generate electro-magnetic radiation it emits and varies with the kinetic energy of the orbiting electron responsible for it.

3.1.4 Potential

Its linear mathematical relationship is: $PE = m.g.R = \frac{1}{2}.m.v^2$

In circular orbits, such as atoms, potential energy between particles is *always* twice the kinetic energy in the orbiting satellite, so their mathematical relationships is:

$PE = CE = 2.\frac{1}{2}.m.v^2 = m.v^2$

At the speed of light this becomes $PE = m.c^2$, which is the true meaning of Henri Poincaré's formula and applies to *orbiting* electrons

3.1.5 Kinetic

Its general mathematical relationship is: $KE = \frac{1}{2}.m.v^2$

The kinetic energy of a satellite orbiting in a circular path is exactly half the satellite's potential energy: $KE = \frac{1}{2}.PE = \frac{1}{2}.m.v^2$

3.2 Orbits

The following Table comprises all the formulas needed to calculate the properties of an orbit.

Sym	Description	units
t	*Orbital period*	*s*
R^P	*Radius at the orbital perigee*	*m*
θ	*Any angle in orbit from apogee*	*°*
R^A	*Radius at the orbital apogee*	*m*
m_2	*Satellite mass*	*kg*
Table 3.2-1: *Input Data*		

Sym	Formula	Description	units
R	p / [1 – e.Cos(θ)]	Orbital radius at θ	m
a	$(R^P + R^A)$ / 2	half the major axis of the ellipse	m
e	$-R^P+\sqrt{[\ R^{P2} - 4.a.(R^P\text{-}a)\ ]}$ / 2.a	eccentricity of the ellipse	
b	$\sqrt{[\ a^2.(1\text{-}e^2)]}$	half the minor axis of the ellipse	m
p	$a.(1\text{-}e^2)$	half-parameter (of orbital path)	m
f	R^P	focus distance (orbital perigee)	m
x'	$a - f$	distance from focus to ellipse centre	m
A	π.a.b	orbital swept area	m²
L	$π.\sqrt{[\ 2.(a^2+b^2) - (a\text{-}b)^2/2.2\ ]}$	orbital path length	m
K	t^2/A^3	orbital constant of proportionality	s²/m³
Table 3.2-2: Orbital Shape			

Sym	Formula	Description	units
m_1	$φ.(2π)^2$ / G.K	Force-centre mass	kg
m_2	*input*	*Satellite mass*	kg
Table 3.2-3: Masses			

Sym	Formula	Description	units
v^P	2.A / t.R^P	satellite velocity at orbital perigee	m/s
v, v_c	2.A / t.R	satellite velocity at θ	m/s
v^A	2.A / t.R^A	satellite velocity at orbital apogee	m/s
g^P	$-v^P.v^A$ / R.(1+e)	gravitational acceleration at perigee	m/s²
g	$-v.v^A$ / R.(1+e)	gravitational acceleration at θ	m/s²
g^A	$-v^P.v^A$ / R.(1+e)	gravitational acceleration at apogee	m/s²
F	$-g.m_2$	gravitational force from force-centre	N
F_c	refer to Chapter 3.2.7	centrifugal force in satellite	N
PE	F/R	potential energy between bodies	J
KE	$½.m_2.v^2$	kinetic energy in satellite	J
E	PE + KE	total energy	J
h	R.v	constant of motion	m²/s
Table 3.2-4: Orbital Performance			

3.2.1 Laws

The following important laws apply to all orbits, irrespective of shape (elliptical or circular):

The key formula for orbital motion originally defined by William Gilbert, and later described as follows by Isaac Newton:
Gravitational Force: $F = G.m_1.m_2 / R^2$
which can be modified to define the following:
Gravitational Energy: $E = G.m_1.m_2 / R$
Gravitational Acceleration: $g = G.m_1/R^2$

Alternative formulas for satellite velocity and gravitational acceleration are:
$v = h/R$
$g = F/m_2$

Throughout any orbit: $E = PE + KE$
where: E is a constant throughout any given orbit
PE is negative and KE is positive

Constant of Proportionality: $K = t^2 / A^2 = (2.\pi)^2 / G.m_1$

3.2.2   Elliptical Orbits

This is nature's perpetual motion machine; it is the first stage of universal energy generation.

A natural (self-energising) orbit is the elliptical path traced by a satellite around its force-centre. Planetary spin (refer to Chapter 3.3) plays no part in this theory.

3.2.2.1 Input Data

Before calculating the properties of an orbit, we must first identify the input data; i.e. the information required to start the calculation. The input information is usually given as follows because it is the easiest to define:

m_2: satellite mass

t: satellite orbital period

R^P: the perigee radial separation (distance) between the centres of mass of the force-centre and its satellites

R^A: the apogee radial separation (distance) between the centres of mass of the force-centre and its satellites
you need the apogee radial separation (distance) for only one satellite in any orbital system
You can calculate this value for all the other orbits about the same force-centre using a constant of proportionality (K).

G: Newton's gravitational constant; 6.67359232004332E-11 m³/kg/s²

3.2.2.2 Orbital Shape

The properties of an ellipse are well known. Its principal dimensions are shown in Fig 23 and described below

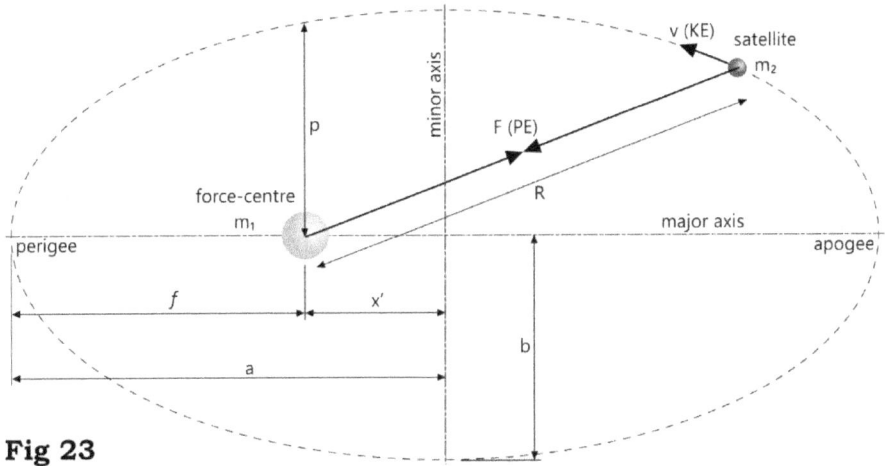

Fig 23

R: radial distance between the satellite and force-centre centres of mass

a: half the major axis of the ellipse

b: half the minor axis of the ellipse

e: eccentricity of the ellipse

p: half-parameter (of orbital path)

f: focus distance from the orbital perigee

x': distance from focus to ellipse centre

v: curvilinear velocity of the satellite

L: circumference of the ellipse (length of orbital path)

t: orbital period; time taken for the satellite to travel around the orbit

θ: angle of 'R' relative to apogee

K: orbital constant of proportionality is the most important orbital feature. It is identical for every orbit encircling the same force-centre

Calculation Procedure

Every orbital system has a unique *constant of proportionality* (K) that is identical for every orbit in the system (a Kepler discovery):
$K = t^2/a^3$
where 't' is the orbital period and 'a' is half the orbit's major axis.

This value for our solar system is:
$K = 2.97491436434708E\text{-}19$ s^2/m^3
and it applies to all the orbits in our solar system.

For the Milky Way galaxy (refer to Chapter 4.2.1)
$K = 3.35025744599744E\text{-}30$ s^2/m^3
and it applies to all the orbits in our galactic system.

First we need to use R^P and R^A for just one of the satellites in an orbital system; e.g. the earth (or our sun), from which, we can obtain a value for 'a':
$a = (R^P + R^A) / 2$
$f = R^P$ and $x' = a - f$
$(R^A = x' + a)$

The constant of proportionality (K) applies to a force-centre and is identical for all of its satellites. All remaining 'a' and 'R^A' values can be determined thus:
$a = \sqrt[3]{[\ t^2/K\ ]}$
$R^A = 2.a - R^P$

We can now calculate orbital eccentricity:
$e = \text{-}R^P + \sqrt{[\ R^{P2} - 4.a.(R^P\text{-}a)\ ]}\ /\ 2.a$

Half its minor axis can be calculated as follows:
$b = \sqrt{[\ a^2.(1\text{-}e^2)]}$

Its half-parameter can be found from either:
$p = a.(1\text{-}e^2)$ or $p/f = 1+e$

The total swept area of the elliptical orbit is calculated as follows:
$A = \pi.a.b$

The total length of the orbital path can be found using:
$L = \pi.\sqrt{[\ 2.(a^2+b^2) - (a\text{-}b)^2/2.2\ ]}$

The radial distance (R) between a satellite and its force-centre at any point in its orbit:

$R = p / [1 - e.Cos(\theta)]$

refer to Fig 8 for a definition of θ

The velocity can now be found anywhere in its orbit from:

$v = 2.A / t.R$

E.g. the greatest and least orbital velocities are calculated as follows:

$v^P = 2.A / t.R^P$ and $v^A = 2.A / t.R^A$

or from:

$(1+e) = v^{P2} / -g^P.R^P$

only applies to the orbital perigee unless the orbit is circular

The gravitational acceleration between a force-centre and its satellite can be found as follows:

$g = -v^P.v^A / R.(1+e)$ {at apogee and perigee}

$g = -v.v^A / R.(1+e)$ {at any other radius}

The centrifugal velocity of a satellite is not the same as its orbital velocity. It must be modified thus:

$a = \sqrt{[\,^4/_3.\pi\,]}$

$\zeta = \sqrt{[\,(f.Sin(\theta/2)^a + p.Cos(\theta/2)^a) / (f.cos(\theta/2)^a + p.Sin(\theta/2)^a)\,]}$

$v_c = \zeta.v$

where 'v' is the orbital velocity as calculated above

Therefore, everything you need to know about an orbit can be found from the nearest and farthest distances of a satellite from its force-centre, without Isaac Newton's gravitational constant, 'G' or the mass of either the force-centre (m_1) or the satellite (m_2).

The force-centre's mass (m_1), which is responsible for its orbital shapes, can now be calculated from the above information.

3.2.2.3 Body Mass

You cannot find a satellite's mass simply from the orbital information described above. In fact, you need to *know* the mass of each satellite in the system in order to determine its performance.

But you can find the mass of the force-centre (m_1), using Isaac Newton's gravitational constant (G) together with the constant of proportionality (K). This means that all the information about every orbit in a given system can be established by using Newton's 'G' only once in one of the orbits.

Calculation Procedure

Force-centre mass may be calculated from the first orbital analysis thus:
$m_1 = (2.\pi)^2 / G.K$
or you may use Isaac Newton's famous formula if you prefer:
$m_1 = -g.R^2 / G$
This is the only time you need to apply Newton's gravitational constant (G); i.e. m_1 (and 'K') remains the same for all other satellite orbits.

You would normally enter a known value for the satellite mass (m_2), but if it is unknown, you can use the following formulas where the gravitational acceleration (g_s) of the satellite at a specified radius (r_s) is known:
either by using Isaac Newton's gravitational constant
$m_2 = -g_s.r_s^2 / G$
or from the following formula if you've forgotten 'G':
$m_2 = m_1 . g_s/g^P . (r_s/R^P)^2$

3.2.2.4 Satellite Performance

Satellite performance constitutes the forces and energies that vary around an orbit. These can be seen in Fig 24 and described below.

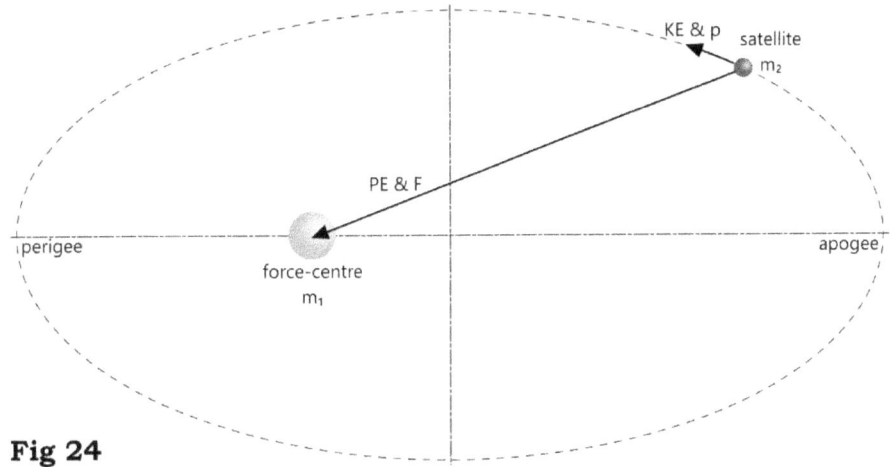

Fig 24

F: gravitational force between the force-centre and the satellite (-ve)

PE: potential energy between the force-centre and the satellite (-ve)

KE: kinetic energy in the satellite (+ve)

E: total energy in the orbital system

p: satellite momentum, which varies with orbital velocity

h: Newton's constant of motion (satellite momentum without the mass component)

Gravitational force may be calculated using Isaac Newton's formula:
$F = -G.m_1.m_2 / R^2$
or an alternative formula:
$F = -g.m_2$

Centrifugal force in the satellite may be calculated as follows:
F_c (refer to Chapter 3.2.7)

Potential energy can be calculated by multiplying the radial distance between the centres of mass of the two bodies by the gravitational force between them:
$PE = F.R$ (where F is always negative)

Kinetic energy in the satellite varies with orbital velocity and can be calculated thus:
$KE = \frac{1}{2}.m_2.v^2$

Total energy is the sum of the two energies above:
$E = PE + KE$

Satellite momentum is calculated thus:
$p = m_2.v$

Newton's constant of motion may be calculated thus:
$h = R.v$
where 'p' is the elliptical half-parameter defined in Chapter 3.2.2.2 above

3.2.3   Circular Orbits

The magnetic potential energy (PE_m) between a force-centre and its satellite(s) applies to all satellites, irrespective of an electrical potential energy (PE) between the same force-centre and its satellite(s).

PE_m is 'φ' times PE

The potential energy between a force-centre and a satellite orbiting in a circular path is *always exactly* twice the satellite's kinetic energy. This means that the potential energy in an electron is; $PE = 2.\frac{1}{2}.m.v^2 = m.v^2$

The principal features of an orbit are described below (and Fig 23):

Minor & Major Axes: are identical

Perigee & Apogee: are identical

R: the distance between the centres of the satellite and its force-centre, which remains constant throughout a satellite's orbit

F: the magnetic or electrical force imposed on the satellite by its force-centre, which remains constant throughout the satellite's orbit

v: the curvilinear velocity of the satellite which remains constant throughout a satellite's orbit

3.2.3.1 Input Data (common)

Before calculating the properties of an orbit, we must first identify the input data; i.e. the information required to start the calculation.

The input information common to both types of circular orbit (magnetic and electrical), is usually given as follows because it is the easiest to define:

e: orbital eccentricity = 0
(for all orbital shells)

G: Newton's gravitational constant; 6.67359232004332E-11 m³/kg/s²
(refer to Chapter 6.11.2)

3.2.3.2 Orbital Shape

The properties of an ellipse are well known. Its principal dimensions are shown in Figs 23 & 25 and described below

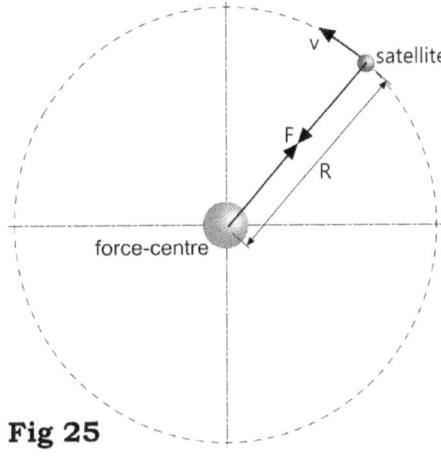

Fig 25

R: radial distance between the satellite and force-centre centres of mass

a = **b** = **p** = f = **R**: orbital radius

e: eccentricity of the ellipse = 0

x': distance from focus to ellipse centre = 0

v: curvilinear velocity of the satellite

L: circumferential length of the circular orbital path

t: orbital period; time taken for the satellite to travel around the orbit

K: orbital constant of proportionality is the most important orbital feature. It is identical for every satellite orbiting the same force-centre

3.2.3.3 Magnetic

The *magnetic* orbit refers to gravitational orbits such as man-made communication satellites.

Input Data (specific)

Before calculating the properties of an orbit, we must first identify the input data; i.e. the information required to start the calculation. The input information is usually given as follows because it is the easiest to define:

m_1: force-centre mass

m_2: satellite mass

v: satellite velocity

Calculation Procedure (Orbital Shape)

First, we need to find the orbital radius, which is defined by the velocity of the satellite:
$R = G.m_1/v^2$

The magnetic attraction holding onto the satellite can be calculated as follows:
$g = v^2/R$

The orbital period is the time taken for a satellite to complete a single orbit:
$t = 2.\pi.R/v$

We now know that in a circular orbit:
$e = 0$ and $a = b = p = f = R$
and the orbital area and the orbital path length are both the same as those for a circle in circular orbits:
$A = \pi.R^2$ and $L = 2.\pi.R$

The constant of proportionality (K) remains constant for all satellites orbiting the same force-centre. It may be calculated thus:
$K = t^2/a^3$
where 't' is the orbital period and 'a' is half the orbit's major axis (R).

m_1 can now be confirmed using 'K' and 'G'.

Body Mass

The satellite and force-centre masses are included in the input data for circular orbit calculations. However, it is useful to know that m_1 can be confirmed as follows:

Calculation Procedure

Force-centre mass may be calculated from the first orbital analysis thus:
$m_1 = (2.\pi)^2 / G.K$

or you may use Isaac Newton's famous formula if you prefer:
$m_1 = g.R^2 / G$

We already know 'm_2' from the input data.

Satellite Performance

Satellite performance constitutes the constant orbital forces and energies. These can be seen in Fig 26 and described below.

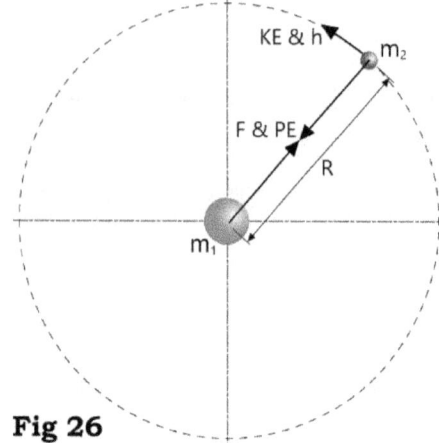

Fig 26

F: attractive force between the force-centre and the Satellite (-ve)

PE: attractive energy between the force-centre and the Satellite (-ve)

KE: kinetic energy in the satellite (+ve)

E: total energy in the orbital system

p: satellite momentum, which varies with orbital velocity

h: Newton's constant of motion (satellite momentum without the mass component)

Calculation Procedure:

Attractive force may be calculated as follows:
$$F = -G.m_1.m_2 / R^2$$
or an alternative formula:
$$F = -g.m_1$$

Centrifugal force in the satellite may be calculated as follows:
$$F_c = m_2.v^2/R$$

Potential energy can be calculated by multiplying the radial distance between the centres of mass of the two bodies by the gravitational force between them:
$$PE = F.R = -2.KE$$

Kinetic energy in the satellite varies with orbital velocity and can be calculated thus:
$$KE = \tfrac{1}{2}.m_2.v^2 = -\tfrac{1}{2}.PE$$

Total energy is the sum of the two energies above:
$$E = PE + KE$$

Satellite momentum is calculated thus:
$$p = m_2.v$$

Newton's constant of motion may be calculated thus:
$$h = R.v$$
where 'p' is the elliptical half-parameter defined in Chapter 3.2.3.2 above

3.2.3.4 Electrical

The *electrical* orbit refers to atomic orbits only.

Input Data (specific)

Before calculating the properties of an orbit, we must first identify the input data relating to atomic particles; i.e. the information required to start the calculation. The input information is usually given as follows because it is the easiest to define:

m_p: proton mass = 1.67262163783E-27 kg

m_e: electron mass = 9.1093897E-31 kg

T: temperature; magnitude of electromagnetic energy (heat) absorbed by the electron

X: Heat constant; 6.9353271647894E-09 K.s^2/m^2
(refer to Chapter 6.6)

X_R: Heat constant; 1.75646616508036E-06 K.m
(refer to Chapter 6.6)

φ: coupling ratio = 4.40742111792335E-40

Calculation Procedure (Orbital Shape)

First, we need the relative electrical charge:
$RAC = e/m_e$

The velocity of an orbiting electron can now be calculated as follows:
$v = \sqrt{[T/X]}$

The orbital radius of an orbiting electron is dependent upon its velocity, and may be calculated thus:
$R = G.m_p / \varphi.v^2$ or $R = X_R/T$

We now know that in a circular orbit:
$e = 0$ and $a = b = p = f = R$

The potential acceleration in the proton-electron pair is:
$g = v^2/R$

The orbital period is the time taken for an electron to complete a single orbit:
$t = 2.\pi.R/v$

We need the separation between any two electrons in an orbit, which will be the same as 2.R. This is calculated for any orbital shell as follows:
$d = \pi . (4.\pi.R^2) / 2.(2.\pi.R)$ {the arc length}
$\ell = 2.R.Sin(d / 2.R)$ {half the straight-line distance}

The orbital area and the orbital path length are both the same as those for a circle in circular orbits:
$A = \pi.R^2$
$L = 2.\pi.R$

An important fact to remember about the constant of proportionality (K) is that it remains constant for all electrons irrespective of shell number (radius) or velocity (temperature).
It may be determined as follows:
$K = t^2/a^3 = 0.15587874533403$ {s^2/m^3}
where 't' is the orbital period and 'a' is half the orbit's major axis.

The above constitutes everything you need to determine the size of every electron shell in an atom. Each shell can hold up to two identical electrons, which is confirmed by 'ℓ' above being the same as 'R'.
m_p can now be confirmed using 'K' and 'G'.

Body Mass

The proton and electron masses are included in the input data for atomic shell calculations. However, it is handy to know that m_p can be confirmed as follows.

Calculation Procedure:

Force-centre mass may be calculated from the first orbital analysis thus:
$m_p = \varphi.(2.\pi)^2 / G.K$

or you may use Isaac Newton's famous formula if you prefer:
$m_p = \varphi.g.R^2 / G$

This is the only time you need to apply Newton's gravitational constant (G); i.e. m_p (and 'K') remains the same for all other electron orbits, irrespective of electron velocity and shell number.

The coupling ratio (φ) is needed in the above formulas because it is the electrical charge that is holding the electron to the proton, not magnetism.

We already know 'm_e' from the input data.

Electron Performance

Electron performance constitutes the constant orbital forces and energies. These can be seen in Fig 27 and described below.

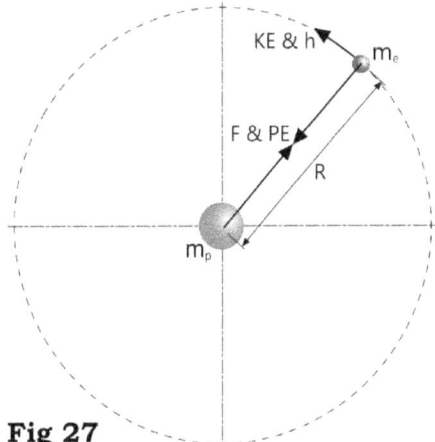

F: attractive force between the proton and the electron (-ve)

PE: attractive energy between the proton and the electron (-ve)

KE: kinetic energy in the electron (+ve)

E: total energy in the orbital system

p: electron momentum, which varies with orbital velocity

Fig 27

h: Newton's constant of motion (electron momentum without the mass component)

Together with the previously calculated information (refer to Chapters 3.2.3.1 to 3.2.3.3), the above provides everything you need to resolve electron performance in atoms.

Calculation Procedure

Electrical attractive force may be calculated as follows:
$$F = -g.m_e$$

Centrifugal force in the orbiting electron may be calculated as follows:
$$F_c = m_e.v^2/R$$

Potential energy can be calculated by multiplying the radial distance between the centres of mass of the two bodies by the gravitational force between them:
$$PE = F.R = 2.KE$$

Kinetic energy in the electron varies with orbital velocity and can be calculated thus:
$$KE = ½.m_e.v^2 = ½.PE$$

Total energy is the sum of the two energies above:
$$E = PE + KE$$

Electron momentum is calculated thus:

$p = m_e.v$

Newton's constant of motion may be calculated thus:

$h = R.v$

You may check the validity of your results as follows:

Calculate the magnetic properties of the atomic orbit using the calculation procedure described in Chapter 3.2.3.3 and compare the calculation results with those in this Chapter (3.2.3.4)

All the following ratios must equal the coupling ratio (φ) ...

g_m/g; F_m/F; Fc_m/Fc; PE_m/PE; KE_m/KE; E_m/E

... and these must equal the square root of the coupling ratio ($\sqrt{\varphi}$):

v_m/v; h_m/h

... if your calculation results are correct.

3.2.4   Force-Centre Mass

As is demonstrated below, it is easy to establish an accurate force-centre mass (m_1) if the major axis ($R_P + R_A$) of just one satellite orbit is known accurately.

$a = \frac{1}{2}.(R_P + R_A)$

Where:

R_P is the radial distance between the centres of the satellite and its force-centre at the satellite's perigee

R_A is the radial distance between the centres of the satellite and its force-centre at the satellite's apogee

$K = t^2/a^3$

Where:

t is the satellite's orbital period

$m_1 = (2.\pi)^2 / G.K$

Where:

G = Newton's gravitational constant

Newton's gravitational constant (G) is not needed for the calculation of any other satellite in the same system (refer to Chapter 3.2.2.2).

3.2.5 Planetary Mass

Using Galileo's measurement of gravitational acceleration, Kepler's law of equal orbital time and swept-Area and Newton's laws of orbital motion we can calculate the mass of a distant satellite from its orbital deviation whilst passing another body of known mass.

By dropping our ball (or rolling it down a slope) and measuring the time it takes to cover a known height, we can find the gravitational acceleration on the surface of our planet, an accurate value for which is now available:

The earth's mean radius: r = 1.496E+11 m
Newton's gravitational constant: G = 6.67359232E-11 m³/kg/s²
The earth's mean gravitational acceleration at sea level:
g = 9.80663139 m/s²

Using Newton's formula; $g = G.m_2/R^2$, we can now find the earth's mass (m_2).

As we know the earth's orbital shape and its velocity (v) at any point within it, we can accurately determine its centrifugal force (F_c) at its perigee for example, from:
$F_c = f/p . m_2.v^2/R$

and because we know that the gravitational force between the earth and sun is exactly the same as the earth's centrifugal force (i.e. they balance each other exactly), we can establish the mass of our sun (m_1) from:
$F_c = G.m_1.m_2 / R^2$
Alternatively, our sun's mass could be determined by using the constant of proportionality from any of its satellite orbits ($m_1 = (2.\pi)^2 / G.K$)

Knowing the mass of our sun and the earth along with the orbital shape of the other bodies in our solar system, we can calculate their masses by comparing force-triangles, one from a planet's theoretical orbit and the other as it passes close to one of which the mass is known, using its orbital deviation and the formulas:
$g = G.m_1/R^2$ & $m_2 = F.R^2 / G.m_1$

So, from measuring the time it takes to drop a ball anywhere on our planet, along with the observed orbits for all the planets and moons in the solar system, Newton has given us all we need to calculate the mass of every celestial body in the solar system.

3.2.6 External Influences

An external influence on a satellite is defined by the gravitational pull between itself and another body.

$$PE = G.m_1.m_2 \, / \, R$$

where:
PE = the potential energy between m_1 and m_2
G = Newton's gravitational constant
m_1 = the mass of the body pulling the satellite out of its orbit
m_2 = the mass of the satellite
R = the radial separation between the centres of the two bodies

3.2.7   Centrifugal Force

Any orbiting mass will be subject to a centrifugal acceleration (a).
Christiaan Huygens gave us the relationship between this and its velocity
$a = v^2 / R$
where 'v' is its curvilinear velocity and 'R' is its radius of motion.

However, the above velocity (v) must be modified for elliptical orbits (v_c) as
shown in Fig 8, and looks like this:

$a = \sqrt{[^4/_3.\pi]}$
$v_c = \zeta.v$

Where: $\zeta = \sqrt{[\ (f.\text{Sin}(\theta/2)^a + p.\text{Cos}(\theta/2)^a) / (f.\text{cos}(\theta/2)^a + p.\text{Sin}(\theta/2)^a)\ ]}$

Centrifugal force, which is equal and opposite to centripetal force, is
calculated thus: $F_c = m_2.v_c^2/R$
which may be simplified at the orbital extremes as follows:
@ the perigee (perihelion) of an ellipse; $F_c = F . f/p = F / (1+e)$
@ the apogee (aphelion) of an ellipse; $F_c = F . p/f = F . (1+e)$

3.2.8 Station-Keeping

The variation in a satellite's PE & CE as it is pulled off course, may be defined by comparing its gravitational (g) and centrifugal (a) acceleration at any given angle through its orbit (θ), by altering its orbital radius.

g = gravitational acceleration (refer to Chapter 3.2.2.2)

a = (v.ʓ)² / R
where:
v = satellite curvilinear velocity (refer to Chapter 3.2.2.2)
ʓ = factor (refer to Chapter 3.2.7)

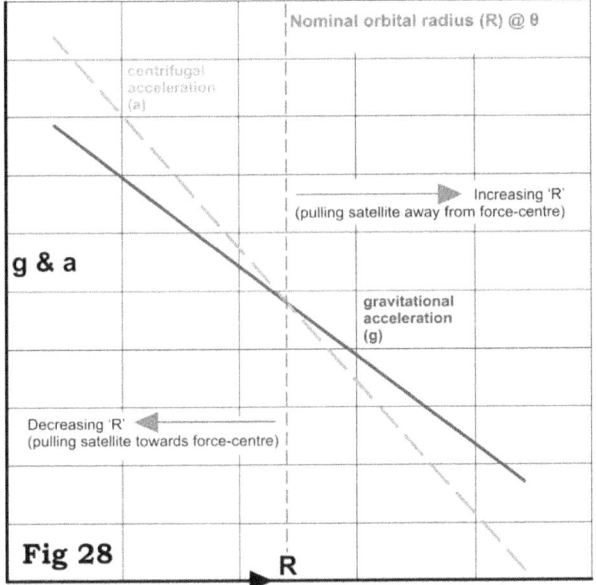

Fig 28

Plotting 'g' and 'a' will result in the following (Fig 28):

As R increases g>a so the satellite will return to its orbit

As R decreases a>g so the satellite will return to its orbit

3.3   Spin

This is nature's energy generator; it is the second stage of universal energy production.

Planetary spin is a fundamental part of the laws of orbital motion but today, it is believed impossible to calculate. Luckily, this is not the case.

Whilst it was not addressed by Isaac Newton - because he did not have the information required to resolve it – it can be solved using his laws of orbital motion.

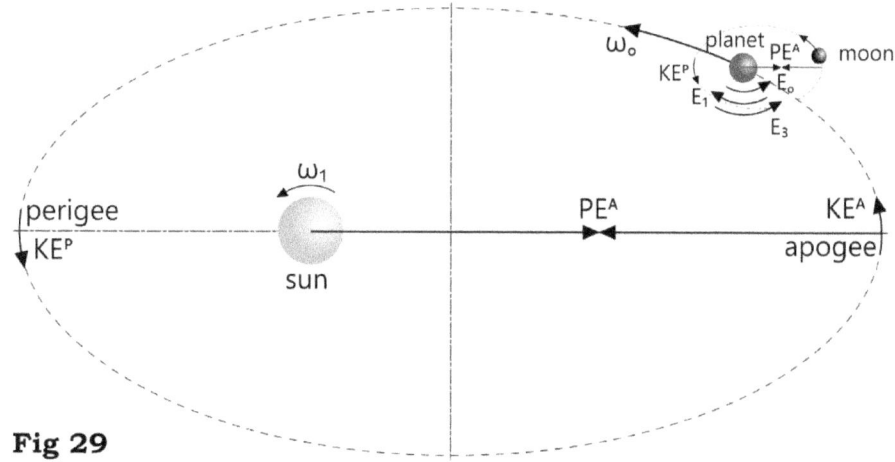

Fig 29

Fig 29 shows the energies that generate spin in each orbiting and/or orbited body. Only orbiting and orbited bodies induce spin in each other. I.e. the sun induces spin in the earth, as does Jupiter and its moons, but Jupiter does not induce spin in Mercury or Neptune.

Spin is induced in orbiting and orbited bodies by the rotational and orbital energies (KE and JE {spin energy}) via the energy that binds them together (PE).

Spin direction: I shall define retrograde as positive and prograde as negative, which defines the orientation of E_0 to E_3 below. This definition may be reversed according to preference.

Spin is induced in one body by the kinetic (orbital and/or rotational) energies in another, via the potential (gravitational) energy between them.

The energies that generate spin in a **satellite** are listed below:

The energy driving a **satellite**'s spin is calculated as follows:
$E_2 = E_1 - E_3 - E_0$

E_0 is the energy developed from the **satellite**'s natural rotation as it orbits around its **force-centre** (star)
$E_0 = \frac{1}{2}.J.\omega_0^2$

E_1 is the rotational energy induced in the **satellite** by the rotational kinetic energy in its **force-centre** (star)
$E_1 = \delta KE . (r/R)^2$
Where; $\delta KE = \Sigma KE^P - \Sigma KE^A$, 'r' is the radius of the **satellite** and 'R' is the radial distance between the centres of the **force-centre** (star) and the **satellite**.

E_3 is the angular velocity induced in a **satellite**'s mass by its orbiting **sub-satellite**(s)
$E_3 = \Sigma(KE^P + PE^A) . Sign[Cos(\theta)]$
Where:
θ is the relative tilt between **satellite** and **sub-satellite** (lunar) orbital planes.
$\Sigma(KE^P + PE^A)$ is the sum of such energies of each **sub-satellite** (moon) orbiting the planet.

The angular velocity of the **satellite** is calculated thus:
$\omega_2 = \sqrt{[2.E_2 / J_2]}$

This calculation method is as accurate as Newton's own laws of motion and is essentially an extension of them. Therefore, not only is planetary spin predictable, it is both simple and accurate, demonstrated by the fact that there now exists a simple calculator for this calculation (http://calqlata.com/proddetail.asp?prod=00085).

Input Data

Before embarking on the calculation procedure, we must first identify the input data, all of which can be found from observation and/or Newton's laws of orbital motion:

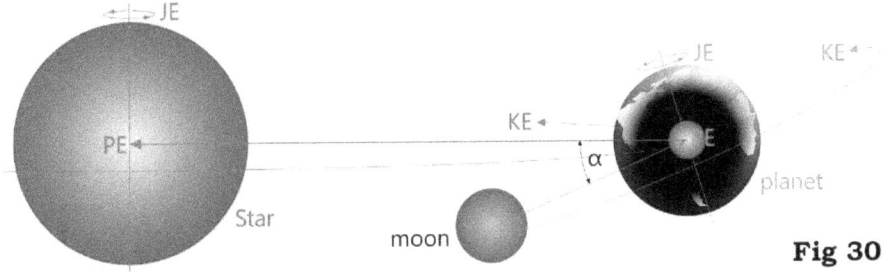

Fig 30

α is the angle between the two orbital planes (Fig 30)

t is orbital period (in seconds). The time taken for the **satellite** to complete one full orbit

r is the outside radius of the **satellite**

Δ is a radial modifier that represents the density variation inside a **satellite**

Δ < 1 means the core density is greater than its surface density
Δ = 1 means that the entire planet is homogeneous
Δ > 1 indicates that the planet is being pulled into a local orbit (e.g. Pluto)
The greater the Δ value (up to 1.0) the lower the density variation.

R^P distance between a **satellite** and its **force-centre** at its orbital perigee

R^A distance between a **satellite** and its **force-centre** at its orbital apogee

KE^P kinetic energy in a **satellite** at its orbital perigee

KE^A kinetic energy in a **satellite** at its orbital apogee

ΣKE^P sum of the kinetic energies in all the **sub-satellites** at their orbital perigee

ΣKE^A sum of the potential energies in all the **sub-satellites** at their orbital apogee

Note: It is important to apply appropriate -ve and +ve polarity to the above 'KE' values according to their prograde or retrograde directions.

Calculation Procedure

This calculation procedure identifies the rotary spin in a **satellite (E_2)** due to its own orbit (**E_0**), its **force-centre (E_1)** and its own **sub-satellites (E_3)**

Spin polarity; defines direction with respect to **sub-satellite** axis tilt:
$\psi = \text{Sign}(\text{Cos}(\alpha))$ {1 or -1}

Average orbital radius of the **satellite**:
$R_{ave} = \frac{1}{2}.(R^P + R^A)$

Polar moment of inertia of the **satellite**:
$J = \frac{2}{5}.m_2.(\Delta.r)^2$

The difference between the kinetic energy in the **satellite** at its orbital perigee and its orbital apogee:
$\delta KE = \Sigma KE^P - \Sigma KE^A$

The sum of all the kinetic energies in the **sub-satellites** at their orbital perigee and their potential energies at their orbital apogee:
$E^s = \Sigma KE^P + \Sigma PE^A$

The rotational velocity in the **satellite** about its axis of spin:
$\omega_0 = 2\pi/t$

The rotational energy in the **satellite** due to 'ω_0':
$E_0 = \frac{1}{2}.J.|\omega_0|\omega_0$
'$|\omega_0|\omega_0$' is the same thing as ω_0^2 multiplied by the sign (polarity; -1 or +1) of ω_0 and therefore gives E_0 direction

The rotational energy in the **satellite** due to the spin in its **force-centre**:
$E_1 = \delta KE.(r/R_{ave})^2$

The total rotational energy in the **satellite**:
$E_2 = E_1 - E_0 - E_3$

The rotational energy in the **satellite** due to its orbiting **sub-satellites**:
$E_3 = \psi.E^s$ (refer to Chapter 4.3.4)

Satellite spin velocity (radians per second) may be calculated as follows:
$\omega_2 = \text{Sign}(E_2) . \sqrt{[2.|E_2| / J]}$
'$|E_2|$' is the positive value of E_2 and 'Sign(E_2)' is its polarity giving ω_2 direction

3.3.1 Polar Moment of Inertia

Polar moment of inertia (J) is calculated thus:

$J = \frac{2}{5}.m.r^2$

where:

'm' is the body's mass

'r' is its outside radius

The radial modifier (Δ) is applied thus:

$J = \frac{2}{5}.m.(Δ.r)^2$ (Δ < 1)

The greater the Δ value the lower the density variation;

Δ < 1 means the core density is greater than its surface density (normal situation)

Δ = 1 means that the entire planet is homogeneous

Δ > 1 indicates that the body is being pulled into a local orbit

3.3.2 Earth's Core

A planet's core will be forced to rotate at a different rate to its outer mantle if the conditions are correct. This is due to the kinetic energy in a large moon inducing rotational energy (E_0) throughout the planet in one direction, and the spin energy in its force-centre (E_1 - E_3) inducing spin in the planet's core in the opposite direction.

When applied to the earth's core, spin theory can show us the difference between the angular velocity of the earth's inner core and its mantle:

Earth's mass; m_E = 5.9662986112E+24 kg
Earth's polar moment of inertia; J_E =1.08209548E+37 = kg.m^2 #
Radius of earth's core; r = 1215000 m
Density of earth's core; ρ = 7870 kg/m^3
Volume of earth's core; V = 7.51307013637E+18 m^3
Mass of earth's core; m = 5.91278619733E+22 kg
Mass of mantle and outer core; m_m = 5.90539190574E+24 kg
Polar Moment of Inertia:
Earth's core; J = $\frac{2}{5}$.m.r^2 = 3.49144112166E+34 kg.m^2
Earth's mantle; J_m = J_E - J = 1.07860404E+37 kg.m^2
Rotational Energies:
E_0 = -2.14473244632E+23 J #
E_1 - E_3 = 2.8770397039E+28 J #
Rotational Velocities:
Earth's core; ω = $\sqrt{[2.E_0/J]}$ = -3.50509019131E-06 ^{c}/s
Earth's mantle; ω_m = sign(E_1+E_3).$\sqrt{[2.(E_1+E_3)/J_m]}$ = 7.3039350764E-05 ^{c}/s
$\delta\omega$ = ω + ω_m = **6.95342605725E-05 ^{c}/s**

Table 4.3.5-1

3.3.3 Earth's Magnetic Field

There are two competing energies driving the spin in the earth's core and its mantle:

$-E_0$ (-2.1447324E+23 J) is the sun's energy driving the core

E_1-E_3 (2.87704E+28 J) is the moon's energy driving the mantle (and core)

The polar moments of inertia:

Core: J = 3.49144112166E+34 kg.m^2

Mantle: J_m = 1.07860404E+37 kg.m^2

The angular velocities:

Core: ω = Sign(E_0) . $\sqrt{[2.|E_0|/J]}$ = -3.50509019131E-06 ^{c}/s

Mantle: ω_m = Sign(E_1-E_3) . $\sqrt{[2.|E_1-E_3|/J_m]}$ = 7.3039350764E-05 ^{c}/s

The differential angular velocity:

$\delta\omega$ = ω + ω_m = **6.95342605725E-05 ^{c}/s**

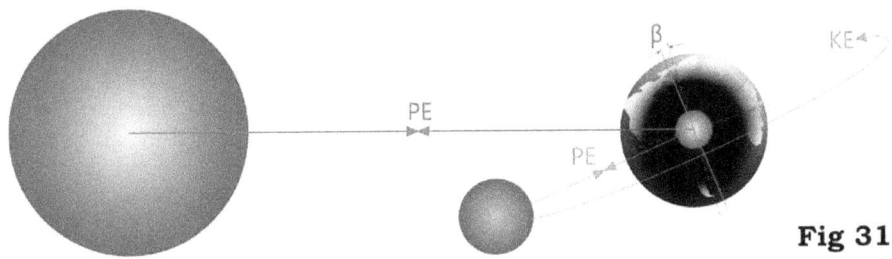

Fig 31

The angular tilt (β) between the two axes can be calculated thus:

β = sign(ω/ω_m) . $\frac{1}{2}.\sqrt{[\ |A\sin(\omega/\omega_m)|\ ]}$

= 0.109553685228394 radians (**6.27696379369167°**)

3.3.4 Magnetic Reversal

N/A

3.3.5 No Moon!

The rotational energy formula (Chapter 5: $E_2 = E_1 - E_3 - E_0$) then becomes:

$E_2 = E_1 - E_0$

In which;

E_1 = 3.208E+23 J

E_0 = 2.14479E+23 J

$E_2 = E_1 - E_0$ = 3.208E+23 - 8.90599E+23 = 1.06321E+23 J

$E_3 = 0$ (zero)

as opposed to 2.87709E+28 J (with its moon)

'*No-moon*' will reduce the earth's internal heat and its magnetic field by a factor of ≈3E+05 ($\delta\omega$≈0).

3.3.6   Hades

What do we know about Hades?

It comprises the same matter as the ultimate-body. It has no force-centre of its own, so it generates no internal friction. It is cold, i.e. it emits insufficient electro-magnetic energy to detect it. It comprises the same matter as all other universal bodies, giving it a density (ρ) of **5300 kg/m³**

If the Milky Way's orbital eccentricity is 0.016 and *if* it takes 230 million years for our solar system to orbit, either or neither of which may be correct, Hades' mass is **1.76572019E+41 kg**

If the above mass and density are correct, Hades' diameter is **4.0E+12 m** from: $r = \sqrt[3]{[\,3.m_1 / 4.\pi.\rho\,]} \approx 1.74966E+12$ m

Hypothetically: *If* we were to define a *photon* as an electron travelling at the speed of light and a black-hole as a black body large enough to trap *photons* (refer to Chapter 6.2.2), then; $PE_{bh} \geq KE_e$

Where; PE_{bh} is the gravitational energy at the surface of a black hole and KE_e is the kinetic energy in a photon:
$PE_{bh} = G.m_1.m_2/R = 6.1350483563E-12$ J
$KE_e = \frac{1}{2}.m_2.c^2 = 4.093555584E-14$ J
Therefore, Hades has more than enough gravitational energy to trap a *photon*.

The minimum size for black-hole comprising mostly iron, may be calculated as follows:
$E = G.\,m_1.\cancel{m_2}/R = \frac{1}{2}.\cancel{m_2}.c^2$
Given that; $R = \sqrt[3]{[3.m_1 / 4.\pi.\rho]}$
$m_1 = \sqrt{[3.c^6 / 32\pi.\rho.G^3]} = $ **9.623785516E+37 kg**
Therefore, Hades is larger than the minimum sized black-hole.

However, as photons do not exist, the above *black-hole* is hypothetical.

If the Milky Way contains 100bn star-systems similar to our solar system, Hades rotates at **2.012413E-07 ᶜ/s**

If the Milky Way contains 10bn star-systems similar to our solar system, Hades rotates at **6.36381E-08 ᶜ/s**

I.e. despite the fact that we cannot see, hear or feel Hades, Newton's laws can tell us an awful lot about it.

Despite the fact that Hades spins on its axis it generates no heat because it has no force-centre of its own.

When calculating the spin in a force-centre that does not follow an orbital path, such as Hades:
Set E_0 & E_1 to zero

When calculating the spin in a sub-satellite that has no sub-sub-satellites, such as our moon:
Set E_3 to zero

Having demonstrated that our sun obeys Newton's laws in orbit around the Milky Way and having established Hades' mass we can now determine its spin rate.

First, however, we need to have a stab at establishing its density. We already know that the largest stable, naturally occurring element in the universe is iron, which is the heaviest atom generated through fusion in the core of a galactic force-centre. We also know that negative magnetic charge (gravity) is 4.407E-40 times that of the positive electrical charge. Using core pressure theory, it has been established that the pressure in the core of Hades is 4E-05 times that required to compress an iron atom and thereby increase its density (refer to Chapter 2.7.2). It is therefore a safe bet that Hades has a density that cannot exceed iron, but is most likely the same as all other universal matter; ≈5300 kg/m³.

Using spin theory, it has been possible to establish the following about Hades:

mass of Hades (m)	1.7657E+41	Kg
density of Hades	≈ 5300	Kg/m³
radius of Hades (r)	≈ 1.996116117E+12	m
moment of angular inertia of Hades ($J_H = \frac{2}{5}.m.r^2$)	2.814190398E+65	kg.m²
energy of the sun used to rotate Hades ($E_S = KE_P - PE_A$)	4.3781615644E+40	J
angular kinetic energy in Hades ($E_H = \frac{1}{2}.J_H.\omega^2$)	4.3781615655E+51	J
number of Suns rotating Hades (N)	1.0E+11 #	
angular velocity of hades ($\omega = \sqrt{[2.E_S.N / J_H]}$)	1.76394139223E-07	°/s
Table 3.3.6-1: Hades Angular Velocity		

 # NASA estimate

You will note, that contrary to popular belief, a *black hole* is simply a slow-spinning ball of matter too cold to emit significant levels of electro-magnetic energy.

When calculating the rotational velocity of Hades, it is important to remember that we know neither the number of star-systems nor do we know the orbital radii of any but our own. Therefore, we can only estimate the total satellite induced spin energy based upon our own solar system multiplied by the number of estimated stars-systems. It is currently estimated that the Milky Way galaxy contains 100bn star-systems. As we have no way of estimating this value with any degree of accuracy, we may try an alternative value to see what effect it would have on Hades' spin-rate. As I believe that 100bn is excessive, I have provided an alternative calculation for 10bn star-systems similar to our own (refer to Chapter 4.3.6)

3.4   Core-Pressure

Core pressure means the internal pressure within any mass (even a lump of metal) due to the same gravitational forces as those attracting satellites to their force-centres.

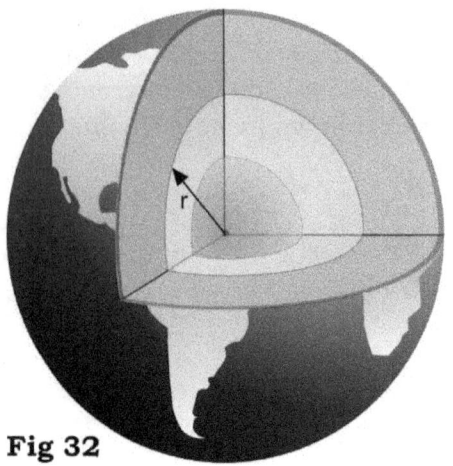

Fig 32

This pressure can be used to define the internal structure of a planet or star, neither of which will be homogeneous. For example, the earth's iron core will have a higher density than its outer core or mantle material. Your structural calculation must generate the same total mass and polar moment of inertia already determined through spin theory (refer to Chapter 3.3).

A simple calculation procedure exists that will provide a fairly reliable result.

The only variable needed when calculating core pressures is a body's 'Δ' value.

The simplest practical calculation procedure is to break the planet up into spherical layers and estimate (iterate) the density and radius of each layer (Fig 32). When the calculation reproduces the correct total mass and polar moment of inertia you will have a representative internal construction.

Newton's gravitational formula $F = G.m_1.m_2 / R^2$ provides all we need to calculate the pressure anywhere inside a body. When used together with the body's Δ value (refer to Chapter 4.3.1), we can determine the internal pressure anywhere inside a body with a complex (layered) structure, such as a planet or star.

But first of all, we need to modify the formula slightly:

$m_2.g = F = G.m_1.m_2 / r^2$

$g = G.m_1 / r^2$

i.e. at any radius within a spherical body, the gravitational acceleration is g based upon the mass of material inside radius r.

Knowing 'g' at any radius 'r', if the total body mass = m

m_1 is the mass inside 'r', and

m_2 is the mass outside 'r' ($m_2 = m - m_1$)

Given that pressure: $p = F/A = m.g / A$

And using the corrected value for Newton's gravitational constant

(G = 8.38628344228057E-10 m³/s²/kg): $p = G.m_2.m_1 / A^2$

For non-homogeneous bodies, they must be broken into layers, each of which will have a representative density. After the final calculation the sum of the polar moment of inertia of each layer must equal the polar moment of inertia of the body ($J = \frac{2}{5}.m.(\Delta.R)^2$) and the sum of the mass of each layer must equal the body's total mass (m). Whilst this is an iterative process it is quite simple and gives an accurate result.

With regard to the calculation for the earth (refer to Chapter 4.4), reducing the core density to that of iron as we understand it, changes the upper mantle density by only a small percentage (<1%). Given the relatively low mass of the earth's core when compared with that of its mantle along with the atomic compressibility of 'φ' (4.407E-40) the core density is almost certainly close to that of iron (7870 kg/m³) and the suggested upper mantle density (close to that of water) must be correct.

This calculation procedure (see below) has revealed that the earth's upper mantle density is significantly less than the crust material it is supporting. It also tells us that the internal pressure is insufficient to support the overlaying crust. However, by applying 'PV=RT', we can establish the amount of heat at the upper-mantle crust interface.

What do we know about the inside of the earth?
1) Its core is made of iron #
2) Its surface comprises mostly water
3) Its inner core radius is ≈1215000m
4) Its outer core radius is ≈3470000m
5) Its upper mantle radius is ≈3463000m *(6km crust and 2km water depth)*
6) Its outer radius is 6371000m
7) Its total mass is 5.96451976771E+24 kg *(Chapter 4.2.2)*
8) Its polar moment of inertia is 1.08209548E+37 kg.m² *(Chapter 4.3.1)*

It is generally believed that the earth's core has a density of 13,000 kg/m³, which is unlikely because the coupling ratio (φ) shows us that the earth's core pressure is insufficient to raise its density to this extent. However, due to the relatively small mass of the iron core (1% of the earth's mass), a significant variation in its density will result in only a minor variation in the calculation result. So, for this exercise, we will set the core density to the popular (but unlikely) density of 13,000 kg/m³

Calculation Procedure

Newton's famous formula may be used to calculate the pressure anywhere within any body (or mass) with one minor modification:

$p = G.m_1.m_2 / A.r^2$

where:

A is the spherical surface area at radius 'r'

m_1 is the mass inside 'r'

m_2 is the mass outside 'r'

Divide the interior up into spherical layers of known (or assumed) radii and assign densities to each layer @ 'r'.

Then iterate through the calculation using the above formula, altering the radii and/or densities until the total mass and polar moment of inertia reflect the known values (input data 7 and 8 above)

CalQlata has created a calculator for core pressure that uses a variable density in each layer. This calculator was used to determine the properties of the internal structure of the earth (refer to Chapter 4.4): (http://calqlata.com/proddetail.asp?prod=00086).

3.5   The Atom

The following calculation procedure provides the formulas required to convert atomic energy to electro-magnetic energy, electricity and the specific heat capacity for any atom.
Refer to Chapter 6.5.2 for a verification of SHC.

There are a few important things to remember when carrying out these calculations:

1) The temperature we feel (and measure) is that emitted by the proton-electron pair(s) with its electron(s) in the inner-most electron shell (T_1). It is this temperature (only) that is used to define the [measured] specific heat capacity of an atom.

2) The specific heat capacity of an atom includes the kinetic energy of all of its electrons. The kinetic energy of the electron(s) in each shell will vary with its orbital radius.

3) All electrons possess the same electrical charge, which means that the spacing between each shell will be equal.

4) You start the calculation procedure by selecting the temperature of the atom; i.e. the temperature that you would feel or measure (T_1)

5) 'n' refers to the electron shell number; 1 to 46 for atomic numbers (Z) 1 to 92 (note: each shell can hold up to two electrons)

Input Data:

measured temperature: T_1

Atomic:

orbital radius of shell 1: $R_1 = X_R/ T_1$
Properties in Shell 'n':
orbital radius: $R_N = R_1.n$
electron temperature: $T_n = X_R/R_n$
electron velocity: $v_n = \sqrt{[T_N/X]}$
kinetic energy of electron: $KE_n = \frac{1}{2}.m_e.v_n^2$
potential energy: $PE_n = -2.KE_n$

Electro-Magnetic:

properties of electro-magnetism radiated by [proton-]electron in shell 'n':
frequency: $f_N = v_n / 2\pi R_n$
wavelength: $\lambda_n = c/f_n$
amplitude: $A_n = R_n$
energy: $E_n = KE_n$
charge: $Q = e'$

Electrical:

current: $A_n = Q.f_n$
voltage: $V_n = E_n/Q$
resistance: $\Omega_n = V_n/A_n$
temperature: $T_n = 2.E_n / k_B.Y$ (will be same as T_N above)

Specific Heat Capacity (SHC):

$SHC = \Sigma KE_n / T_1.m.Y$ {J / kg.K}

Where:
m = atomic mass

3.5.1 Atomic Particles

The only two particles in the universe necessary to make it work are the proton and the electron. All the calculations in this section of the book concern these two particles alone.

Sub-atomic particles are not considered here as they are unnecessary.

3.5.1.1 The Electron

The mass of an electron: m_e = 9.1093897E-31 kg (constant)

The electrical charge of an electron never changes, irrespective of its situation:

e = -1.60217648753E-19 C (constant)

The relationship between temperature (T), velocity (v) and orbital radius (R) may be defined as follows:
v^2 = T/X and R = X_R/T
where: X & X_R are heat transfer coefficients (refer to Chapter 6.6)

It is the electro-static *potential* energy that holds an orbiting electron to its proton. This energy is *always* twice the kinetic energy in the electron:
KE = ½.m_e.v^2
PE = 2.KE = 2.½.m_e.v^2 = m_e.v^2

At the time of ejection from its proton-electron pair:

Angular velocity in an electron is: ω = 2π / orbital period
(at the time of ejection)

The linear velocity of an electron is: v = $\sqrt{[2.KE / m]}$
(at the time of ejection)

3.5.1.2 The Proton

The mass of a proton: m_p = 1.67262163783E-27 kg

The electrical charge of a lone proton, that is not part of proton-electron pair is the same as that of an electron:

e = -e = 1.60217648753E-19 C (constant)

After acquiring an orbiting electron, it will have the ability to top up its electrical charge to a maximum:

e' = -e.m_p/m_e = 2.94183820093364E-16 C

3.5.1.3 The Neutron

This is nature's energy storage facility; it is the third and last stage of universal energy production.

$RAC = k_B.R_i.Q_e = 96485.3317942156$ C/mol (of electrons)
Note: Faraday's constant = 96485.3317942158 C/mol {exact}

Rest condition @ $T = 1K$: $N_t = 1$ $N_v = 1.5$ $N_p = 2.5$

$RAM_p = R_i.m_p/k_B = 1.00727638277235$ N
Note: $RAM_H = 1.00794$ g/mol (hydrogen)
$RAM_e = RAM_p . m_e/m_p = 0.000548580318390698$ g/mol
$R_a = RAM_e / R_i = 15156.3563034308$ J/g/K
$R = R_a.m_p = 1.38065156E\text{-}23$ J/K
$k_B = 1.38065156E\text{-}23$ J/K
$k_B.N_A.L_n(N_t) = c_p.L_n(T).RAM_e = 3.371231032$ J/K/mol
$exp(2.5xL_n(T)) = 1$

$c_v = N_t.R_a = 22734.5344551462$ J/g/K
$C_v = m_e.c_v = 2.07097734E\text{-}23$ J/K
$c_p = c_v+R_a = 37890.8907585769$ J/g/K
$C_p = m_p.c_v = 3.4516289E\text{-}23$ J/K

$KE_e = k_B.T.N_p = 3.4516289E\text{-}23$ J

$X = T_n.t_n^2/(2\pi.R_n)^2 = T_n/c^2 = 6.9353271647894E\text{-}09$ K.s²/m²
$T = X.v^2$ K

Or @ the speed of light; when a neutron is created:
$T_n = X.c^2 = 623316124.71718$ K

According to Newton's orbital motion formula; $v = \sqrt{[R.g]}$
'v' will be the speed of light (c)
when: R_n = orbital radius of 1.46677550700177 x $(r_p + r_e)$
$R_n = 2.817937953839E\text{-}15m$
$g = G.m_p / \varphi.R_n^2 = 3.18940728807829E\text{+}31 m/s^2$

The magnetic field 'B' as described by Lorentz ($R_n = 1E\text{-}07 . e/B$), is actually 1/RC, where RC is the relative charge capacity of Quanta (1.75882E+11 C/kg). And the orbital radius at which an electron and a proton combine to create a neutron may be calculated as follows:
$R_n = \mu'.e.RC = 2.817937953839E\text{-}15$ m

Where: 'e' is the elementary charge unit (1.6021764875E-19 C)
According to the heat-coefficient 'X', this orbital radius occurs when the electron is travelling at the speed of light (Table 4.5.3-4).

At the orbital radius; 'R_n', the attractive magnetic force exceeds the repulsive electrical force and the electron combines with its proton to create a neutron, which occurs when the electro-magnetic energy is equivalent to a temperature of 623316124.71718 K
Moreover, 'R_n' occurs when: $KE = \tfrac{1}{2}.m.c^2$
Note: In circular orbits; $PE = -2.KE = -2 . \tfrac{1}{2}.m.c^2 = -m.c^2$

The magnetic constant (μ_o), which controls this union between an electron and a proton, is referred to as;
$\mu_o = 1E\text{-}07 . 4\pi$ H/m,
but what exactly is 1E-07 and what is a *Henry*?
$m_e.R_n/e^2 = 1E\text{-}07$ (*exactly*) kg.m/C^2
$\mu_o = 4.\pi.R_n.m_e/e^2$ kg.m/C^2
and; Henry $= $ kg.m^2/C^2

Verifying that '$R_{n'}$ is a real and important physical constant (that I refer to as the **neutronic radius**) and that it occurs when $PE = m.c^2$

This is the rationale behind $E = m.c^2$

Moreover, all of the above can *only* be determined using Newton's laws of motion and Coulomb's electrical force. None of this is achievable with Relativity, in which mass is claimed to vary with velocity and gravity (magnetic charge) is claimed to deform the orbital path.

A neutron therefore has no electrical charge but possesses a magnetic charge and its magnitude is that of a proton plus that of an electron. It also stores the following energy:

$E = KE - PE = mc^2 = -4.093555611312680E\text{-}14$ J

It cannot be mere coincidence that Newton's and Coulomb's laws show that an electron orbiting its proton at the speed of light (c) comes within striking distance of each other, and that this occurs at a temperature of; ≈6.23E+08 K, or that Newton's and Coulomb's laws can explain all of the above clearly and accurately, but Relativity cannot.

3.5.2   Electron Shells

The innermost orbital radius is calculated thus; $R_1 = X_R/T$
then add this radius (R_1) to each subsequent shell radius (R_2 to R_n)

The electron temperature in subsequent shells can then be determined using the same formula: $T = X_R/R$

The orbital velocity of each electron is determined thus: $v = \sqrt{[T/X]}$

Their kinetic energies are: $KE = \frac{1}{2}.m.v^2$
Their potential energies are: $PE = -2.KE$

Electro-magnetic energy is equal to KE.

You use the formulas provided in Chapter 3.1.3 to determine the properties of the electro-magnetic energy generated by each proton-electron pair.

Johannes Rydberg (refer to Chapter 3.1.3) gave us another useful constant (R_γ) that enables us to calculate electron shell number (n) from an electron's kinetic energy:

$n = (KE_n/R_\gamma) . (a_0/R_n) = (c/v_e)^2$

Where; 'v_e' is the velocity of the electron

3.5.3   Nucleus

Atomic lattice structure and density are defined thus; $\zeta = 3\sqrt{[\,m.n/\rho\,]} / R$
The resultant factors are listed below for Hydrogen to Uranium[4]:

Hydrogen	(1)	4.25	Silver	(47)	5.625
Helium	**(2)**	**1.5**	Cadmium	(48)	3.875
Lithium	(3)	19.5	Indium	(49)	6.125
Beryllium	(4)	16.25	Tin	(50)	6.875
Boron	(5)	17.5	Antimony	(51)	5.625
Carbon	(6)	18	Tellurium	(52)	4.75
Nitrogen	(7)	5.5	Iodine	(53)	3
Oxygen	(8)	5	**Xenon**	**(54)**	**2**
Fluorine	(9)	4.625	Caesium	(55)	6
Neon	**(10)**	**1.625**	Barium	(56)	6.875
Sodium	(11)	11.125	Lanthanum	(57)	8
Magnesium	(12)	10	Cerium	(58)	6.625
Aluminium	(13)	11.875	Praseodymium	(59)	7.875
Silicon	(14)	14	Neodymium	(60)	7.25
Phosphorus	(15)	6	Promethium	(61)	6.375
Sulphur	(16)	6.5	Samarium	(62)	5.5
Chlorine	(17)	6.5	Europium	(63)	5.875
Argon	**(18)**	**2.375**	Gadolinium	(64)	7
Potassium	(19)	10	Terbium	(65)	6.875
Calcium	(20)	10.75	Dysprosium	(66)	6.125
Scandium	(21)	12	Holmium	(67)	6.125
Titanium	(22)	10.375	Erbium	(68)	6.25
Vanadium	(23)	9.5	Thulium	(69)	5.25
Chromium	(24)	8	Ytterbium	(70)	4.625
Manganese	(25)	6.875	Lutetium	(71)	6.5
Iron	(26)	7.875	Hafnium	(72)	6.75
Cobalt	(27)	7.5	Tantalum	(73)	6.75
Nickel	(28)	7.5	Tungsten	(74)	6.5
Copper	(29)	7	Rhenium	(75)	6.375
Zinc	(30)	4.875	Osmium	(76)	5.875
Gallium	(31)	7.375	Iridium	(77)	5.5
Germanium	(32)	8.5	Platinum	(78)	5.25
Arsenic	(33)	4.375	Gold	(79)	4.75
Selenium	(34)	4.875	Mercury	(80)	2.375
Bromine	(35)	5.375	Thallium	(81)	4.125
Krypton	**(36)**	**2.25**	Lead	(82)	4.5
Rubidium	(37)	7	Bismuth	(83)	4.5
Strontium	(38)	7.625	Polonium	(84)	3.75
Yttrium	(39)	9.5	Astatine	(85)	3
Zirconium	(40)	9.5	**Radon**	**(86)**	**2**
Niobium	(41)	8.875	Francium	(87)	4.75
Molybdenum	(42)	8.25	Radium	(88)	5.625
Technetium	(43)	7.625	Actinium	(89)	6
Ruthenium	(44)	7.25	Thorium	(90)	7
Rhodium	(45)	6.875	Protactinium	(91)	5.75
Palladium	(46)	6.25	Uranium	(92)	5.5

Note; the lowest values for ζ occur at **the noble gases**.

Isotopic factors ($\Gamma = 9.[\psi-1]$) are listed below, for Hydrogen to Uranium[4]:

Element		Value	Element		Value
Hydrogen	(1)	0.07146	Silver	(47)	2.655612766
Helium	**(2)**	**0.011709 (0)**	Cadmium	(48)	3.0770625
Lithium	(3)	2.823	Indium	(49)	3.089020408
Beryllium	(4)	2.2774095	Tin	(50)	3.3642
Boron	(5)	1.45998	Antimony	(51)	3.485294118
Carbon	(6)	0.01605	Tellurium	(52)	4.084615385
Nitrogen	(7)	0.008614286	Iodine	(53)	3.54981566
Oxygen	(8)	0.0001125	**Xenon**	**(54)**	**3.984 (4)**
Fluorine	**(9)**	**0.998403 (1)**	Caesium	(55)	3.748015636
Neon	(10)	0.16173	Barium	(56)	4.070410714
Sodium	(11)	0.809811818	Lanthanum	(57)	3.932447368
Magnesium	(12)	0.22875	Cerium	(58)	3.742137931
Aluminium	(13)	0.679526308	Praseodymium	(59)	3.494387288
Silicon	(14)	0.054964286	Neodymium	(60)	3.636
Phosphorus	(15)	0.5842566	Promethium	(61)	3.393442623
Sulphur	(16)	0.0365625	Samarium	(62)	3.826451613
Chlorine	(17)	0.769235294	Europium	(63)	3.709142857
Argon	**(18)**	**1.974 (2)**	Gadolinium	(64)	4.11328125
Potassium	(19)	0.520247368	Terbium	(65)	4.005047077
Calcium	(20)	0.0351	Dysprosium	(66)	4.159090909
Scandium	(21)	1.266818571	Holmium	(67)	4.154819104
Titanium	(22)	1.581954545	Erbium	(68)	4.137220588
Vanadium	(23)	1.933630435	Thulium	(69)	4.034896957
Chromium	(24)	1.4985375	Ytterbium	(70)	4.248
Manganese	(25)	1.77769764	Lutetium	(71)	4.178915493
Iron	(26)	1.331653846	Hafnium	(72)	4.31125
Cobalt	(27)	1.6444	Tantalum	(73)	4.308645205
Nickel	(28)	0.865735714	Tungsten	(74)	4.360135135
Copper	(29)	1.721172414	Rhenium	(75)	4.34484
Zinc	(30)	1.6155	Osmium	(76)	4.527236842
Gallium	(31)	2.24216129	Iridium	(77)	4.466922078
Germanium	(32)	2.4215625	Platinum	(78)	4.509
Arsenic	(33)	2.433163636	Gold	(79)	4.439227215
Selenium	(34)	2.901176471	Mercury	(80)	4.566375
Bromine	(35)	2.546742857	Thallium	(81)	4.709255556
Krypton	**(36)**	**2.9495 (3)**	Lead	(82)	4.741463415
Rubidium	(37)	2.789464865	Bismuth	(83)	4.660523133
Strontium	(38)	2.752105263	Polonium	(84)	4.390928571
Yttrium	(39)	2.516734615	Astatine	(85)	4.233917647
Zirconium	(40)	2.5254	**Radon**	**(86)**	**5.024407 (5)**
Niobium	(41)	2.394083415	Francium	(87)	5.071034483
Molybdenum	(42)	2.558571429	Radium	(88)	5.116193182
Technetium	(43)	2.701297674	Actinium	(89)	4.95788764
Ruthenium	(44)	2.673409091	Thorium	(90)	5.20381
Rhodium	(45)	2.5811	Protactinium	(91)	5.142857143
Palladium	(46)	2.821304348	Uranium	(92)	5.285436848

Note; Γ = 0 to 5' occur at **the noble gases** (except Neon).

3.5.4 How They Work

An electron absorbs electro-magnetic energy and converts it into velocity. This means that at the 'speed-of-light' its kinetic energy reaches:
$KE = \frac{1}{2}.m_e.c^2$
and the coincident *potential* energy between the proton and its electron is:
$PE = m_e.c^2$

There are three key energy conditions for the proton-electron pair according to Planck, each of which relate to electron velocities, shell radii and associated temperatures. These energies are listed below:

Neutronic: $KE_n = \frac{1}{2}.m_e.c^2 = 4.0935556113127E\text{-}14$ J

Mean: $KE_m = \frac{1}{2}.m_e.v_m^2 = 2.3771466644364E\text{-}17$ J

Minimum: $KE_o = \frac{1}{2}.m_e.v_o^2 = 1.3804200555196E\text{-}20$ J

There is also a fourth energy level that appears to be when an orbiting electron may leave its proton partner and continue in free-flight. This condition is referred to as the 'cold' energy level:

Cold: $KE_c = \frac{1}{2}.m_e.v_c^2 = 8.0161630672150E\text{-}24$ J

$\delta = KE_n / KE_c = 5.1066271693683E\text{+}09$

The potential energy between a proton and its orbiting electron is twice the kinetic energy of the electron.

The energy stored within each neutron is therefore: $-KE_n = PE_n + KE_n$

The orbiting electron defines the properties of the electro-magnetic energy emitted by the proton-electron pair (refer to Chapter 3.1.3)

3.5.4.1 Planck Electron Velocities

Neutronic: c = 299792459 m/s

Mean: $v_m = \sqrt{[T_m/X]}$ = 7224342.80705 m/s

Minimum: $v_o = \sqrt{[T_o/X]}$ = 174090.866621 m/s

Cold: $v_c = \sqrt{[T_c/X]}$ = 4195.2092599072 m/s

This means that an electron is unlikely to exceed (or even achieve) the *speed of light* in free-flight unless it is provided with artificial energy greater than $\frac{1}{2}.m.c^2$ or knocked from its orbit, and most electrons in free-flight will be travelling at little more than 4000m/s

3.5.4.2 Planck Shell Radii

Neutronic: $R_n = X_R/T_n = 2.81793795383896E\text{-}15$ m

Mean: $R_m = X_R/T_m = 4.85261843362263E\text{-}12$ m

Minimum: $R_o = X_R/T_o = 8.35643156381572E\text{-}09$ m

Cold: $R_c = X_R/T_c = 1.43901585166681E\text{-}05$ m [2]

3.5.4.3 Planck Temperatures

Neutronic: $T_n = X.c^2 = 623316124.71718$ K

Mean: $T_m = X.v_m^2 = 361962.55467156$ K

Minimum: $T_o = X.v_o^2 = 210.19332853584$ K

Cold: $T_c = X.v_c^2 = 0.122060237421696$ K

The maximum possible natural temperature cannot exceed; T_n

4 Calculation Results

A collection of [mostly] tabulated calculation results for selected examples using the formulas provided in section 3 (above).

4.1   Energy

The calculation results for energy can be found Chapters 4.2 to 4.5 under the symbols PE & KE

4.1.1 Electricity

Refer to Chapter 6.9

4.1.2 Magnetism

Refer to Chapters 6.7 & 6.8

4.1.3   Electro-Magnetic Energy

Tables -1 to -2 below show the properties of electro-magnetic radiation emitted by a proton-electron pair at Planck's temperatures (refer to Chapter 3.5.4.3).

	minimum (T_o)	neutronic (T_n)
$\underline{T}$ (K)	210.19332853584	623316124.71718
f (Hz)	3.31570021944219E+12	1.69320448260835E+22
λ (m)	9.04160325599144E-05	1.77056263481051E-14
A (m)	8.3564315638157E-09	2.817937953839E-15 #
Spectra	Infra-red	γ

Table 4.1.3-1: Planck's Temperature Boundaries

$\underline{T}$ = temperature; f = frequency; λ = wavelength; A = amplitude
the neutronic radius (R_n) is 2.81793795383896E-15 m

	cold (T_c)	mean (T_m)
T_c (K)	0.122060237421696	361962.55467156
f (Hz)	46398953.1627279	2.3694215485103E+17
λ (m)	6.46119014686784	1.26525589837944E-09
A (m)	1.43901585166681E-05	4.85261843362268E-12
Spectra	radio	X

Table 4.1.3-2: Planck Temperatures

$\underline{T}$ = temperature; f = frequency; λ = wavelength; A = amplitude

We may deduce from the above Tables that:

The lowest possible electro-magnetic energy is that associated with a temperature of 0.12206 K and;

The highest possible electro-magnetic energy is that associated with a temperature of 623316124.71718 K

4.2 Orbits

The following Tables comprise the calculation results from Chapter 3.2 for the orbital systems with which we are most familiar:
Galactic: Milky Way
Solar: Solar System
Lunar: Those in our solar system
Atomic: Proton-electron pair

Tables -1: *Input Data*

G = *Newton's gravitational constant*
t = *orbital period*
R = *radial distance between the centres of the force-centre and the satellite*
P = *perigee &* A = *apogee*
r = *radius of* m_2
m_2 = *satellite mass*

Tables -2: Orbital Shape

a & b = major and minor orbital semi-axes
e = orbital eccentricity
p = orbital half parameter
f = distance between orbital 'focus' and satellite (R^P)
x' = distance between orbital 'focus' and orbit centre $(a - f)$
A = total swept area of orbit
L = circumferential length of orbit
K = constant of proportionality
v = satellite curvilinear velocity
v_c = centrifugal velocity at 'θ'
g = gravitational acceleration between force-centre and satellite

---

Tables -3: Masses

m_1 = force-centre mass
m_2 = *satellite mass*

Tables -4: Orbital Performance

F = gravitational force between force-centre and satellite
PE = gravitational energy between force-centre and satellite
KE = kinetic energy in satellite
E = total energy should always be the same, irrespective of radial distance
h = constant of motion should always be the same, irrespective of radial distance

Suffixes: A = @ apogee; P = @ perigee; none = @ R (θ)

4.2.1 Galactic

The following Tables show the orbital properties of our sun based upon two galactic population options;
1) 100bn star-systems, which is the popularly held belief, but appears far too high to me, and;
2) 10bn star-systems, which also seems too high

The reason for performing both calculations is to show that irrespective of the number of star-systems in the Milky Way, the orbital calculations can be performed successfully without the need for dark matter, moreover the number of star-systems orbiting Hades has no effect on its orbital shapes or velocities.

You will notice that the results in both calculation options are exactly the same, despite the difference in star-system populations. This is because:

1) Hades' mass *only* defines the orbital radii and period, and
2) Hades' spin-rate is *only* defined by star-system population

It is difficult to understand therefore, how anybody could have declared 100 years ago that Newton's laws of orbital motion would predict the ejection of the Milky Way's stars into outer space through centrifugal force when:

3) Newton's laws of orbital motion are independent of star-system population
4) The presence of a galactic force-centre was not predicted at that time
5) The mathematical laws of planetary spin have only just been discovered

Note: All three values for total energy (E) should be identical if the input data is correct. The small difference at θ simply means that NASA's input data is slightly out.

Property	The Sun (10bn)	The Sun (100bn)	units
t	7.25825E+15	7.25825E+15	s
R^P	2.4653729E+20	2.4653729E+20	m
$R (\theta = 45°)$	2.533231129E+20	2.533231129E+20	m
R^A	2.5452510E+20	2.5452510E+20	m
m_2	1.9885E+30	1.9885E+30	kg

Table 4.2.1-1: *Input Data*

Property	The Sun (10bn)	The Sun (100bn)	units
a	2.505312E+20	2.505312E+20	m
e	0.015941744	0.015941744	
b	2.504994E+20	2.504994E+20	m
p	2.504675E+20	2.504675E+20	m
f	2.465373E+20	2.465373E+20	m
x'	3.993904E+18	3.993904E+18	m
A	1.971598E+41	1.971598E+41	m²
L	1.574034E+21	1.574034E+21	m
K	3.3502574E-30	3.3502574E-30	s²/m³
v^P	220360.56213	220360.56213	m/s
v	214457.71443	214457.71443	m/s
vc	215677.54520	215677.54520	m/s
v^A	213444.94588	213444.94588	m/s
g^P	-1.93872549E-10	-1.93872549E-10	m/s²
g	-1.83625048E-10	-1.83625048E-10	m/s²
g^A	-1.81894819E-10	-1.81894819E-10	m/s²

Table 4.2.1-2: Orbital Shape

Property	The Sun (10bn)	The Sun (100bn)	units
m_1	**Hades: 1.76572E+41**	**Hades: 1.76572E+41**	kg
m_2	1.9885E+30	1.9885E+30	kg

Table 4.2.1-3: Masses

Property	The Sun (10bn)	The Sun (100bn)	units
F^P	-3.8551556E+20	-3.8551556E+20	N
Fc^P	3.8551556E+20	3.8551556E+20	N
PE^P	-9.5043962E+40	-9.5043962E+40	J
KE^P	4.8279564E+40	4.8279564E+40	J
E^P	**-4.6764398E+40**	**-4.6764398E+40**	J
h^P	5.4327096E+25	5.4327096E+25	m²/s
F	-3.6513841E+20	-3.6513841E+20	N
Fc	3.6514104E+20	3.6514104E+20	N
PE	-9.2497998E+40	-9.2497998E+40	J
KE	4.5727657E+40	4.5727657E+40	J
E	**-4.6770342E+40**	**-4.6770342E+40**	J
h	5.4327096E+25	5.4327096E+25	m²/s
F^A	-3.6169785E+20	-3.6169785E+20	N
Fc^A	3.6160593E+20	3.6160593E+20	N
PE^A	-9.2061180E+40	-9.2061180E+40	J
KE^A	4.5296782E+40	4.5296782E+40	J
E^A	**-4.6764398E+40**	**-4.6764398E+40**	J
h^A	5.4327096E+25	5.4327096E+25	m²/s

Table 4.2.1-4: Orbital Performance

4.2.2 Solar

A solar system is simply a collection of planets and comets orbiting a star. Our solar system is known to us and therefore enables us to accurately identify the orbital properties of any and every member in great detail.

Whilst 21 solar orbits have been analysed in the creation of this book, I have chosen the results from just three to include here:

Tables -1 to -4 below show the orbital properties of Mercury, Earth and Jupiter, calculated using the formulas provided in Chapter 3.2.2

The 21 satellites analysed included; three of the largest asteroids (Ceres, Pallas & Vesta), three of its outermost planets (Eris, Haumea & MakeMake) and four of its comets (Halley's, C/2001 OG108 Loneos, 3200 Phaeton & P/2009 WX51 Catalina (CSS)). Whilst they have been analysed with similar accuracy to the planets, as their input data is suspect, their orbital output data is considered (by me) to be too unreliable to warrant inclusion here.

Note: All three values for total energy (E) should be identical if the input data is correct. The small difference at θ simply means that the input data is slightly out.

Property	Mercury	Earth	Jupiter	units
t	7600521.6	31558118.4	374335689.6	s
R^P	4.60012E+10	1.47095E+11	7.4052E+11	m
$R\ (\theta = 45°)$	6.489479E+10	1.520941962E+11	8.036758283E+11	m
R^A	6.981450E+10	1.5209420E+11	8.156104331E+11	m
m_2	2439700	5.9645198E+24	1.89819E+27	kg

Table 4.2.2-1: *Input Data*

Property	Mercury	Earth	Jupiter	units
a	5.790785E+10	1.49595E+11	7.78065E+11	m
e	0.205613749	0.016709147	0.048254588	
b	5.667055E+10	1.49574E+11	7.77159E+11	m
p	5.545968E+10	1.49553E+11	7.76253E+11	m
f	4.60012E+10	1.47095E+11	7.4052E+11	m
x'	1.190665E+10	2499598078	37545216557	m
A	1.030967E+22	7.02945E+22	1.89966E+24	m²
L	3.599701E+11	9.39865E+11	4.88588E+12	m
K	2.974914E-19	2.9749144E-19	2.974914E-19	s²/m³
v^P	58974.212404	30286.0088	13705.90205	m/s
v	41804.350699	29436.3192	12628.84142	m/s
v_c	44699.545376	29611.76945	12844.06876	m/s
v^A	38858.468158	29290.5356	12444.04703	m/s
g^P	-0.062711465	-0.0061332328	-2.419979E-04	m/s²
g	-0.0315112486	-0.0057939183	-2.054582E-04	m/s²
g^A	-0.0272266369	-0.0057366715	-1.994894E-04	m/s²

Table 4.2.2-2: Orbital Shape

Property	Mercury	Earth	Jupiter	units
m_1	1.9885E+30	1.9885E+30	1.9885E+30	kg
m_2	3.3011E+23	5.96451976771E+24	1.89819E+27	kg

Table 4.2.2-3: Masses

Property	Mercury	Earth	Jupiter	units
F^P	-2.0701682E+22	-3.6581678E+22	-4.589200E+23	N
Fc^P	2.070168E+22	3.6581788E+22	4.593581E+23	N
PE^P	-9.523022E+32	-5.3809820E+33	-3.398394E+35	J
KE^P	5.740543E+32	2.7354550E+33	1.782892E+35	J
E^P	**-3.782479E+32**	**-2.6455270E+33**	**-1.615503E+35**	J
h^P	2.712885E+15	4.4549138E+15	1.014465E+16	m²/s
F	-1.040218E+22	-3.4557837E+22	-3.896267E+23	N
F_c	1.016378E+22	3.4557932E+22	3.896404E+23	N
PE	-6.750472E+32	-5.2300158E+33	-3.131336E+35	J
KE	2.884507E+32	2.5841189E+33	1.513689E+35	J
E	**-3.865964E+32**	**-2.6458968E+33**	**-1.617647E+35**	J
h	2.712885E+15	4.4549138E+15	1.014465E+16	m²/s
F^A	-8.987785E+21	-3.4216388E+22	-3.783076E+23	N
Fc^A	8.607808E+21	3.4206938E+22	3.777870E+23	N
PE^A	-6.274777E+32	-5.2041141E+33	-3.085516E+35	J
KE^A	2.492298E+32	2.5585865E+33	1.469714E+35	J
E^A	**-3.782479E+32**	**-2.6455275E+33**	**-1.615802E+35**	J
h^A	2.712885E+15	4.4549138E+15	1.014465E+16	m²/s

Table 4.2.2-4: Orbital Performance

4.2.3 Lunar

A lunar system is simply a collection of moons orbiting a planet, just as planets orbit their stars. There is no limit to the number of moons any given planet can host. Whilst our own planet (Earth) can claim only one, Jupiter and Saturn host upwards of 100 between them. In fact, there are probably over 200 moons in our solar system, 152 of which have been analysed in the creation of this book.

Tables -1 to -4 below show the orbital properties of three of the best-known; our own moon, Phobos and Titan, each of which were calculated using the formulas provided in Chapter 4

Whilst it is by no means definitive, moons orbiting closer to their planet tend to orbit in a prograde direction and those further away tend to orbit in a retrograde direction.

The orbital planes of the moons in our solar system indicates that they have most probably come from outside our solar system. Jupiter's lunar orbital planes are only 3° from Jupiter's own orbital plane because of the strong influence of spin between the solar system's two most massive bodies.

It is most likely that planets orbiting closest the solar force-centre will not collect moons, because the sun's gravitational energy along with its proximity will be sufficient to trap or deflect most, if not all, of them from these planets. This is why Mercury and Venus are the only planets in our solar system without moons, and why our Earth has only one, whilst much smaller planets such as Mars and Pluto have managed to trap many more.

Note: All three values for total energy (E) should be identical if the input data is correct. The small difference at θ simply means that the input data is slightly out.

Property	Moon	Phobos	Titan	units
t	2360620.8	27553.84387	1378079.998	s
R^P	359508000	9230818.576	1.186590E+09	m
$R\ (\theta = 45°)$	398868087.4	9473266.673	1.246426E+09	m
R^A	406504000	9516150	1.257304E+09	m
m_2	1737494.514	11166.66667	2574707.348	kg

Table 4.2.3-1: *Input Data*

Property	Moon	Phobos	Titan	units
a	383006000	9373484.288	1221946914	m
e	0.061351519	0.015220137	0.0289348727	
b	382284501.4	9372398.529	1221435284.29	m
p	381564362	9371312.896	1220923868	m
f	359508000	9230818.576	1186590036	m
x'	23498000	142665.712	35356878.42	m
A	4.599834E+17	2.759953E+14	4.688918E+18	m²
L	2.404232E+09	5.889193E+07	7.676112E+09	m
K	9.918265E-14	9.218520E-13	1.040859E-15	s²/m³
v^P	1084.020134	2170.247491	5734.922357	m/s
v	977.0496126	2114.704625	5459.6124	m/s
v_c	998.1341195	2126.191187	5515.7452	m/s
v^A	958.6963727	2105.174976	5412.376679	m/s
g^P	-0.0030796892	-0.5025949109	-0.026938071	m/s²
g	-0.0025018740	-0.4771984141	-0.024413781	m/s²
g^A	-0.0024087646	-0.4729072361	-0.02399316	m/s²

Table 4.2.3-2: Orbital Shape

Property	Moon	Phobos	Titan	units
m_1	5.964367E+24	6.417101E+23	5.683400E+26	kg
m_2	7.346377E+22	1.065853E+16	1.345525E+23	kg

Table 4.2.3-3: Masses

Property	Moon	Phobos	Titan	units
F^P	-2.2624559E+20	-5.3569229E+15	-3.624585E+21	N
Fc^P	2.2624559E+20	5.3569229E+15	3.624585E+21	N
PE^P	-8.1337101E+28	-4.9448783E+22	-4.300896E+30	J
KE^P	4.3163628E+28	2.5100700E+22	2.212671E+30	J
E^P	**-3.8173473E+28**	**-2.4348083E+22**	**-2.088225E+30**	J
h^P	3.8971391E+11	2.0033161E+10	6.805002E+12	m²/s
F	-1.8379710E+20	-5.0862336E+15	-3.284935E+21	N
F_c	1.8349395E+20	5.0863023E+15	3.284231E+21	N
PE	-7.3310799E+28	-4.8183247E+22	-4.094428E+30	J
KE	3.5065212E+28	2.3832343E+22	2.005328E+30	J
E	**-3.8245587E+28**	**-2.4350904E+22**	**-2.089100E+30**	J
h	3.8971391E+11	2.0033161E+10	6.805002E+12	m²/s
F^A	-1.7695694E+20	-5.0404959E+15	-3.228340E+21	N
Fc^A	1.7629087E+20	5.0393283E+15	3.225637E+21	N
PE^A	-7.1933704E+28	-4.7966115E+22	-4.059004E+30	J
KE^A	3.3760231E+28	2.3618032E+22	1.970778E+30	J
E^A	**-3.8173473E+28**	**-2.4348083E+22**	**-2.088225E+30**	J
h^A	3.8971391E+11	2.0033161E+10	6.805002E+12	m²/s

Table 4.2.3-4: Orbital Performance

4.2.4 Atomic

All electrons in an atom must obey Newton's laws of orbital motion, and their spacing between adjacent electrons in the same and adjacent shells is maintained by Coulomb's force-law. These conditions define the amount of electro-magnetic energy any given electron is able to absorb.

Each shell can hold up to two identical electrons, both of which will absorb the same amount of electro-magnetic energy (heat or temperature) from their surroundings.

This means that the electro-magnetic energy (heat or temperature) absorbed in each shell will be different from each of the other shells. Moreover, in accordance with Newton's laws of motion, electron(s) in the innermost shell will absorb the most energy and the outermost shell will absorb the least.

The heat we feel from matter is the sum of the electro-magnetic energy radiated by each proton-electron pairing within an atom. Its temperature is that generated by the proton-electron pairs in the innermost shell.

The Tables in this Chapter list electron orbital performance at the maximum possible temperature; when the orbiting electron achieves *light-speed.*

Below is listed the descriptions of the additional symbols used in the *electron* Tables.

Tables -1 to -4: (specific to the atom)

T = the temperature of the electron
X = velocity heat coefficient [5]
X_R = radial heat coefficient [5]

Tables -1 to -4 below show the orbital properties of a proton-electron pair at the maximum possible temperature (when the neutron is created).

Sym	Newton	Coulomb	units
T	623316124.7171790	623316124.7171790	K
m_2	9.1093897E-31	9.1093897E-31	kg
X		6.9353271647894E-09	$K.s^2/m^2$
X_R		1.75646616508036E-06	$K.m$

Table 4.2.4-1: _Input Data_ (T_n)

Sym	Newton	Coulomb	units
R	2.817937953839E-15	2.817937953839E-15	m
d	8.852813174052E-15	8.852813174052E-15	m
ℓ	2.817937953839E-15	2.817937953839E-15	m
a	2.817937953839E-15	2.817937953839E-15	m
e	0	0	
b	2.817937953839E-15	2.817937953839E-15	m
p	2.817937953839E-15	2.817937953839E-15	m
f	2.817937953839E-15	2.817937953839E-15	m
x'	0	0	m
A	2.494667824141E-29	2.494667824141E-29	m^2
L	1.770562634810E-14	1.770562634810E-14	m
K	0.15587874533403	0.15587874533403	s^2/m^3

Table 4.2.4-2: Orbital Shape (T_n)

Sym	Newton	Coulomb	units
m_1	1.67262163783E-27	1.6726216378300E-27	kg
m_2	9.1093897E-31	9.1093897E-31	kg

Table 4.2.4-3: Masses (T_n)

Sym	Newton	Coulomb	units
v	6.2938005855237E-12	299792459	m/s
g	1.4057061035135E-08	3.189407288078E+31	m/s^2
F	1.2805124700573E-38	29.05355389912620	N
Fc	1.2805124700573E-38	29.05355389912620	N
PE	-3.6084046897386E-53	-8.1871112226254E-14	J
KE	1.8042023448693E-53	4.0935556113127E-14	J
E	-1.8042023448693E-53	-4.0935556113127E-14	J
h	1.7735539543841E-26	8.4479654849081E-07	m^2/s
PE/KE	-2	-2	

Table 4.2.4-4: Orbital Performance (T_n)

4.3 Spin

Below is listed the descriptions of the symbols used in Chapters 4.3.4 to 4.3.6.

θ = planetary tilt
ψ = Sign(Cos(θ))
Δ = radial modifier for J (density variable)
J = polar moment of inertia
R_{ave} = average orbital radius
ω_0 = rotational velocity due to orbit
E_0 = natural rotational energy due to orbit
E_1 = rotational energy induced by its force-centre
E_2 = total rotational energy
E_3 = rotational energy induced by its sub-satellites (e.g. moons)
ω_2 = total rotational velocity

4.3.1   Polar Moment of Inertia

Below is listed the polar moment of inertia and radial modifier for each of the planets in our solar system:

Planet	J {kg.m^2}	Δ
Our Sun	3.90008074E+46	0.318
Mercury	5.19308435E+35	0.813
Venus	3.30912713E+37	0.681
Earth	1.08209548E+37	0.334
Mars	1.58326892E+31	0.00232
Jupiter	1.92585538E+39	0.0228
Saturn	1.52392272E+38	0.0141
Uranus	1.38906233E+37	0.0249
Neptune	1.05696850E+38	0.0652
Pluto	5.48499917E+35	8.642

4.3.2   The Earth's Core

$\delta\omega = \omega + \omega_m =$ **6.95342605725E-05** ^{c}**/s**

I.e. the earth's mantle is rotating 6.9534E-05 radians per second faster than its inner core, which is responsible for creating the earth's internal heat (through friction) and its magnetic field.

The positive value for '$\delta\omega$' shows the correct rotational direction according to the right-hand-rule (North is currently at the top of planet).

4.3.3   The Earth's Magnetic Field

The angular tilt (β) between the two axes can be calculated thus:

$\beta = \text{sign}(\omega/\omega_m) \cdot \frac{1}{2}.\sqrt{[\ |A\sin(\omega/\omega_m)|\ ]}$

$= 0.109553685228394$ radians (**6.27696379369167°**)

4.3.4 Our Sun

The following Table lists the spin characteristics of our sun based upon two different galactic populations. Note: they are identical.

Sym	Sun (100bn)	Sun (10bn)	units
θ	45	45	°
ψ	1	1	
Δ	0.318284697814735	0.318284697814735	
J	3.90008074E+46	3.90008074E+46	kg.m^2
R_{ave}	2.50531194E+20	2.50531194E+20	m
ω_0	8.65661425E-16	8.65661425E-16	°/s
E_0	1.46130117E+16	1.46130117E+16	J
E_1	2.30014080E+16	2.30014080E+16	J
E_2	1.60100474E+35	1.60100474E+35	J
E_3	-1.60100474E+35	-1.60100474E+35	J
ω_2	**2.86532908E-06**	**2.86532908E-06**	°/s

Table 4.3.4-1: Spin Velocity of Our Sun (star-systems)

4.3.5   Our Planets

Planetary spin has been analysed for all the planets within our solar system and is listed in Tables -1 to -3 below.

Sym	Mercury	Venus	Earth	units
θ	0.01	177.4	23.4	°
ψ	1	-1	1	
Δ	0.812862196423	0.68123190998	0.3342776983	
J	5.19308435E+35	3.30912713E+37	1.08209548E+37	kg.m²
R_{ave}	5.79078502E+10	1.08205784E+11	1.49594598E+11	m
ω_0	8.26678173E-07	3.23643522E-07	1.99098857E-07	°/s
E_0	1.77446862E+23	1.73307475E+24	2.14473245E+23	J
E_1	5.76562924E+23	2.51533292E+23	3.20799628E+23	J
E_2	3.99116062E+23	-1.48154146E+24	2.87701826E+28	J
E_3	0	0	-2.87700762E+28	J
ω_2	**1.23980080E-06**	**-2.99236920E-07**	**7.29211510E-05**	°/s

Table 4.3.5-1: Spin Velocity in the Inner Planets

Sym	Mars	Jupiter	Saturn	units
θ	25.2	3.1	26.7	°
ψ	1	1	1	
Δ	**0.0023170868178**	0.022780669614	0.014060010927	
J	1.58326892E+31	1.92585538E+39	1.52392272E+38	kg.m²
R_{ave}	2.27934435E+11	7.78065217E+11	1.42682770E+12	m
ω_0	1.05858984E-07	1.67848952E-08	6.75904500E-09	°/s
E_0	8.87115436E+16	2.71288224E+23	3.48099680E+21	J
E_1	1.55613559E+22	2.52841886E+26	9.19174032E+24	J
E_2	3.97741612E+22	2.97776807E+31	2.04408655E+30	J
E_3	-2.42128940E+22	-2.97774281E+31	-2.04407737E+30	J
ω_2	**7.08823600E-05**	**1.75852520E-04**	**1.63788410E-04**	°/s

Table 4.3.5-2: Spin Velocity in the Middle Planets

Sym	Uranus	Neptune	Pluto	units
θ	97.8	28.3	122.5	°
ψ	-1	1	-1	
Δ	0.024937619324	0.06523792741	8.64241984998	
J	1.38906233E+37	1.05696850E+38	5.48499917E+35	kg.m²
R_{ave}	2.86949539E+12	4.49637396E+12	5.90394017E+12	m
ω_0	2.36992355E-09	1.20822828E-09	8.03026194E-10	°/s
E_0	3.90086044E+19	7.71489543E+19	1.76850379E+17	J
E_1	2.80792156E+22	2.09359675E+21	6.27106256E+15	J
E_2	-7.1182956E+28	6.20291286E+29	-3.55514898E+25	J
E_3	7.11829844E+28	-6.20291284E+29	3.55514896E+25	J
ω_2	**-1.0123767E-04**	**1.0833825E-04**	**-1.1385592E-05**	°/s

Table 4.3.5-3: Spin Velocity in the Outer Planets

4.3.6   Hades

The following Table lists the spin characteristics of Hades based upon two different galactic populations; note: they are different.

Sym	Hades (100bn)	Hades (10bn)	units
θ	0	0	°
ψ	1	1	
Δ	1	1	
J	2.8141903980768E+65	2.8141903980768E+65	kg.m^2
R_{ave}	0	0	m
ω_0	0	0	°/s
E_0	0	0	J
E_1	0	0	J
E_2	4.378161564734E+51	4.3781615647344E+50	J
E_3	-4.378161564734E+51	-4.3781615647344E+50	J
ω_2	**1.76394139222867E-07**	**5.57807245849103E-08**	°/s

Table 4.3.6-1: Spin Velocity of Hades (star-systems)

4.4   Core Pressure

A core pressure calculation using the above theory (refer to Chapter 3.4) for the earth is presented in Fig 33 in which it can be seen that the upper mantle material has a density of $1105 kg/m^3$.

A description of the input and output data from CalQlata's program (*Cores*) for pressures within for our earth is listed below:

R_0 to R_6 = outside radii for each spherical layer
ρ_0 to ρ_6 = density at each spherical radius
m = mass of the planet
J = polar moment of area of the planet
p_6 = atmospheric pressure at the planet's outer surface
G = Newton's gravitational constant
Fm = mass-factor (must equal 1)
FJ = polar moment of area-factor (must equal 1)
p_0 to p_5 = pressure at each spherical radius (R_0 to R_6)

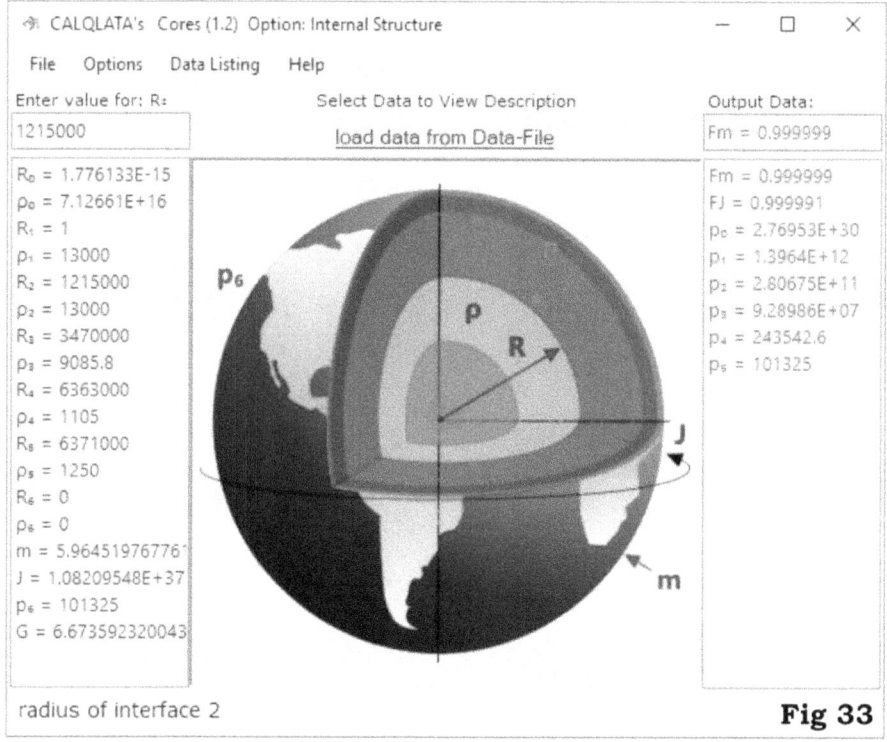

▲ CALQLATA's Cores (1.2) Option: Internal Structure   — ☐ ✕

File   Options   Data Listing   Help

Enter value for: R: Select Data to View Description   Output Data:

1215000   load data from Data-File   Fm = 0.999999

$R_0 = 1.776133E-15$
$\rho_0 = 7.12661E+16$
$R_1 = 1$
$\rho_1 = 13000$
$R_2 = 1215000$
$\rho_2 = 13000$
$R_3 = 3470000$
$\rho_3 = 9085.8$
$R_4 = 6363000$
$\rho_4 = 1105$
$R_5 = 6371000$
$\rho_5 = 1250$
$R_6 = 0$
$\rho_6 = 0$
$m = 5.96451976776$
$J = 1.08209548E+37$
$p_6 = 101325$
$G = 6.673592320043$

Output Data:

$Fm = 0.999999$
$FJ = 0.999991$
$p_0 = 2.76953E+30$
$p_1 = 1.3964E+12$
$p_2 = 2.80675E+11$
$p_3 = 9.28986E+07$
$p_4 = 243542.6$
$p_5 = 101325$

radius of interface 2

Fig 33

With kind permission of CalQlata

4.5   The atom

The following lists the symbol descriptions for Tables -1 to -4:

Tables -1: *Input Data*

T = electron temperature
m_2 = electron mass
X = Heat coefficient for electron velocity and temperature
X_R = Heat coefficient for electron orbital radius and temperature

---

Tables -2: Orbital Shape

R = radial distance between the centres of the force-centre and the satellite
d = arc distance between equi-spaced points on the surface of a sphere
ℓ = straight-line distance between equi-spaced points on the surface of a sphere
a & b = major and minor orbital semi-axes
e = orbital eccentricity
p = orbital half parameter
f = distance between orbital 'focus' and satellite (R^p)
x' = distance between orbital 'focus' and orbit centre (a - f)
A = total swept area of orbit
L = circumferential length of orbit
K = constant of proportionality

Tables -3: Masses

m_1 = force-centre mass
m_2 = satellite mass

Tables -4: Orbital Performance

v = satellite curvilinear velocity
g = gravitational acceleration between force-centre and satellite
F = gravitational force between force-centre and satellite
F_c = centrifugal force on electron
PE = gravitational energy between force-centre and satellite
KE = kinetic energy in satellite
E = total energy should always be the same, irrespective of radial distance
h = constant of motion should always be the same, irrespective of radial distance
PE/KE = confirmation of 2:1 relationship in circular orbits

4.5.1   Cold (T_c)

Tables -1 to -4 below show the orbital properties of a proton-electron pair at the minimum temperature.

Property	Newton	Coulomb	units
T	0.122060237421696	0.122060237421696	K
m_2	9.1093897E-31	9.1093897E-31	kg
X		6.9353271647894E-09	$K.s^2/m^2$
X_R		1.75646616508036E-06	$K.m$

Table 4.5.1-1: *Input Data* (T_c)

Property	Newton	Coulomb	units
R	1.439015851667E-05	1.439015851667E-05	m
d	1.439015851667E-05	1.439015851667E-05	m
ℓ	4.520801627996E-05	4.520801627996E-05	m
a	1.439015851667E-05	1.439015851667E-05	m
e	0	0	
b	1.439015851667E-05	1.439015851667E-05	m
p	1.439015851667E-05	1.439015851667E-05	m
f	1.439015851667E-05	1.439015851667E-05	m
x'	0	0	m
A	6.505505204927E-10	6.505505204927E-10	m^2
L	9.041603255991450E-05	9.041603255991450E-05	m
K	0.1558787453340300	0.1558787453340300	s^2/m^3

Table 4.5.1-2: Orbital Shape (T_c)

Property	Newton	Coulomb	units
m_1	1.67262163783E-27	1.6726216378300E-27	kg
m_2	9.1093897E-31	9.1093897E-31	kg

Table 4.5.1-3: Masses (T_c)

Property	Newton	Coulomb	units
v	8.8073631286363E-17	4195.2092599071500	m/s
g	5.3904649618566E-28	1.223042867389E+12	m/s^2
F	4.9103846001748E-58	1.1141174098854E-18	N
Fc	4.9103846001748E-58	1.1141174098854E-18	N
PE	-7.0661212774321E-63	-1.6032326134430E-23	J
KE	3.5330606387160E-63	8.0161630672150E-24	J
E	-3.5330606387160E-63	-8.0161630672150E-24	J
h	1.2673935153494E-21	6.0369726260658E-02	m^2/s
PE/KE	-2	-2	

Table 4.5.1-4: Orbital Performance (T_c)

4.5.2   Planck Minimum ($\underline{T}_o$)

Tables -1 to -4 below show the orbital properties of a proton-electron pair at Planck's minimum temperature [3].

Property	Newton	Coulomb	units
T	210.19332853584	210.19332853584	K
m_2	9.1093897E-31	9.1093897E-31	kg
X		6.9353271647894E-09	$K.s^2/m^2$
X_R		1.75646616508036E-06	$K.m$

Table 4.5.2-1: *Input Data* ($\underline{T}_o$)

Property	Newton	Coulomb	units
R	8.356431563816E-09	8.356431563816E-09	m
d	2.625250401111E-08	2.625250401111E-08	m
ℓ	8.356431563816E-09	8.356431563816E-09	m
a	8.356431563816E-09	8.356431563816E-09	m
e	0	0	
b	8.356431563816E-09	8.356431563816E-09	m
p	8.356431563816E-09	8.356431563816E-09	m
f	8.356431563816E-09	8.356431563816E-09	m
x'	0	0	m
A	7.396196699443E-10	7.396196699443E-10	m^2
L	9.640713088869E-05	9.640713088869E-05	m
K	0.15587874533403	0.15587874533403	s^2/m^3

Table 4.5.2-2: Orbital Shape ($\underline{T}_o$)

Property	Newton	Coulomb	units
m_1	1.67262163783E-27	1.6726216378300E-27	kg
m_2	9.1093897E-31	9.1093897E-31	kg

Table 4.5.2-3: Masses ($\underline{T}_o$)

Property	Newton	Coulomb	units
v	3.6548390907795E-15	174090.86662108400	m/s
g	1.5985111201450E-21	3.6268626876711E+18	m/s^2
F	1.4561460733184E-51	3.3038505610385E-12	N
Fc	1.4561460733184E-51	3.3038505610385E-12	N
PE	-1.2168185008604E-59	-2.7608401110393E-20	J
KE	6.0840925043021E-60	1.3804200555196E-20	J
E	-6.0840925043021E-60	-1.3804200555196E-20	J
h	3.0541412738857E-23	1.4547784128045E-03	m^2/s
PE/KE	-2	-2	

Table 4.5.2-4: Orbital Performance ($\underline{T}_o$)

4.5.3 Planck Mean ($\underline{T}_m$)

Tables -1 to -4 below show the orbital properties of a proton-electron pair at Planck's mean temperature.

Property	Newton	Coulomb	units
T	361962.55467156	361962.55467156	K
m_2	9.1093897E-31	9.1093897E-31	kg
X		6.9353271647894E-09	$K.s^2/m^2$
X_R		1.75646616508036E-06	$K.m$

Table 4.5.3-1: Input Data ($\underline{T}_m$)

Property	Newton	Coulomb	units
R	4.852618433623E-12	4.852618433623E-12	m
d	1.524495042174E-11	1.524495042174E-11	m
ℓ	4.852618433623E-12	4.852618433623E-12	m
a	4.852618433623E-12	4.852618433623E-12	m
e	0	0	
b	4.852618433623E-12	4.852618433623E-12	m
p	4.852618433623E-12	4.852618433623E-12	m
f	4.852618433623E-12	4.852618433623E-12	m
x'	0	0	m
A	2.193772531476E-16	2.193772531476E-16	m²
L	5.250500802222E-08	5.250500802222E-08	m
K	0.15587874533403	0.15587874533403	s²/m³

Table 4.5.3-2: Orbital Shape ($\underline{T}_m$)

Property	Newton	Coulomb	units
m_1	1.67262163783E-27	1.6726216378300E-27	kg
m_2	9.1093897E-31	9.1093897E-31	kg

Table 4.5.3-3: Masses ($\underline{T}_m$)

Property	Newton	Coulomb	units
v	1.5166683358448E-13	7224342.80705005	m/s
g	.7402920143405E-15	1.0755250944965E+25	m/s²
F	4.3181167250426E-45	9.7973772178982E-06	N
Fc	4.3181167250426E-45	9.7973772178982E-06	N
PE	-2.0954172818476E-56	-4.7542933288727E-17	J
KE	1.0477086409238E-56	2.3771466644364E-17	J
E	-1.0477086409238E-56	-2.3771466644364E-17	J
h	7.3598127242123E-25	3.5056979076300E-05	m²/s
PE/KE	-2	-2	

Table 4.5.3-4: Orbital Performance ($\underline{T}_m$)

4.5.4 Neutron (T_n)

Tables -1 to -4 below show the orbital properties of a proton-electron pair at the maximum possible temperature (when the neutron is created).

Property	Newton	Coulomb	units
T	623316124.71718	623316124.71718	K
m_2	9.1093897E-31	9.1093897E-31	kg
X		6.9353271647894E-09	$K.s^2/m^2$
X_R		1.75646616508036E-06	$K.m$

Table 4.5.4-1: *Input Data* (T_n)

Property	Newton	Coulomb	units
R	2.817937953839E-15	2.817937953839E-15	m
d	8.852813174052E-15	8.852813174052E-15	m
ℓ	2.817937953839E-15	2.817937953839E-15	m
a	2.817937953839E-15	2.817937953839E-15	m
e	0	0	
b	2.817937953839E-15	2.817937953839E-15	m
p	2.817937953839E-15	2.817937953839E-15	m
f	2.817937953839E-15	2.817937953839E-15	m
x'	0	0	m
A	2.494667824141E-29	2.494667824141E-29	m^2
L	1.770562634810E-14	1.770562634810E-14	m
K	0.15587874533403	0.15587874533403	s^2/m^3

Table 4.5.4-2: Orbital Shape (T_n)

Property	Newton	Coulomb	units
m_1	1.67262163783E-27	1.6726216378300E-27	kg
m_2	9.1093897E-31	9.1093897E-31	kg

Table 4.5.4-3: Masses (T_n)

Property	Newton	Coulomb	units
v	6.2938005855237E-12	299792459	m/s
g	1.4057061035135E-08	3.189407288078E+31	m/s^2
F	1.2805124700573E-38	29.05355389912620	N
Fc	1.2805124700573E-38	29.05355389912620	N
PE	-3.6084046897386E-53	-8.1871112226254E-14	J
KE	1.8042023448693E-53	4.0935556113127E-14	J
E	-1.8042023448693E-53	-4.0935556113127E-14	J
h	1.7735539543841E-26	8.4479654849081E-07	m^2/s
PE/KE	-2	-2	

Table 4.5.4-4: Orbital Performance (T_n)

The following Table is a check-list of the ratios between various Newton (^{N}) and Coulomb (^{C}) calculated properties:

Property Ratio		Value
$v^N : v^C$	$\sqrt{\varphi}$	2.0993858906650E-20
$g^N : g^C$	φ	4.40742111792335E-40
$F^N : F^C$	φ	4.40742111792335E-40
$Fc^N : Fc^C$	φ	4.40742111792335E-40
$PE^N : PE^C$	φ	4.40742111792335E-40
$KE^N : KE^C$	φ	4.40742111792335E-40
$E^N : E^C$	φ	4.40742111792335E-40
$h^N : h^C$	$\sqrt{\varphi}$	2.09938589066502E-20

All of which are either the coupling ratio or its square-root

Appendices

References, symbols, glossary, etc. used throughout this book along with a summary list of corollaries and hypotheses.

A1 General

N/A

A2 References

Most of the references used for the creation of this book are from original work supplied in CalQlata (www.calqlata.com), but some of additional sources are listed below:

Magnificent Principia; Colin Pask; 978-1-61614-745-7

Seven Brief Lessons on Physics; Carlo Rovelli; 978-0-141-98172-7

Science Data Book; Open University; 0 05 002487 6

Science and Technology Dictionary; Chambers; 0-550-18026-5

A Dictionary of Scientific Units; H G Jerrard & D B McNeill; 0-412-28100-7

It is important to note here that most of the sources here are from work done by pre-20th Century scientists that are universally known and available from sources too numerous to mention here.

A3 Glossary

Atomic Particle	One of the three components that comprise an atom
Big-Bang	The eruption that occurred when the Ultimate-Body accumulated sufficient 'mass' to compromise the integrity of the innermost neutron.
Black-Body	A collection of Quanta too cold to emit electro-magnetic radiation in the frequencies that would enable detection.
Coupling Ratio (φ)	The ratio of the coupling force due to a magnetic charge to the coupling force due to an electric charge: $\varphi = G.m_p.m_e \div k.e^2$ (refer to Chapter 6.11.3)
Gas	Atoms that possess greater electrical field energy than magnetic field energy
Hades	The Milky Way's force-centre
Proton-electron pair	A proton that hosts an orbiting electron
Sub-Atomic Particle	The many particles said to compromise atomic particles (leptons, gluons, fermions, quarks, etc.)
Ultimate-Body	A body that contains all the Quanta in the universe ($\approx 2.8E+75$) and represents the maximum single 'mass' that can exist without generating a Big-Bang.
Ultimate Density	The mass-density of all three atomic particles $\rho = 7.12660796350449E+16$ kg/m^3 Nothing in nature has a 'mass'-density greater than this value
Universal Period	The time elapsed since the last Big-Bang or between subsequent Big-Bangs
Viscous	Solid or liquid matter in which magnetic field energy is greater than electrical field energy

All other definitions can be found on the following web page:
http://calqlata.com/help_definitions.html

A4 Symbols

Refer to Chapter 5 for a list of all the symbols used in this book.

The most prominent subscripts are listed below:

mass	e	electron
	p	proton
temperature	c	cold
	o	minimum Planck
	m	mean Planck
	n	maximum Planck
Rydberg	γ	energy constant
	∞	wave number
	o	orbital radius (a_o) *occasionally referred to as the Bohr radius*
Others	u	Ultimate
radii	n	Neutron orbit radius
	s	Schwarzschild radius
	1	force-centre
	2	satellite
energy	e	electron
	p	proton

The most prominent superscripts are listed below:

Force	N	Newton
	P	Planck

A5 Useful Formulas

Equidistant arc-length between 'n' points on the surface of a sphere:
$d = \pi.A / C.n$
where C is the circumference of the sphere
Linear distance across arc-length 'd' (above):
$\ell = 2.R.Sin(\frac{1}{2}.d/R)$
but if you know 'ℓ' and need to find 'n':
$n = \pi / Asin(\frac{1}{2}.\ell/R)$
and if $\ell=R$:
$n = \pi / Asin(\frac{1}{2}) = \mathbf{6}$

Lorentz's Equation (magnetic force or field strength):
$F = q.v.B$
Which becomes:
$F = q.g.R.B$
for the laws of orbital motion
Where:
q is the total electrical charge $= q_1.q_2 / m_e.(q_1+q_2)$
v = relative velocity (electrical circuits)
g = gravitational attraction between m_1 & m_2
R = radial separation between m_1 & m_2
$B = \mu_o.I / 2.\pi.R$ kg/C {$R = 2.R_n$}
$I = e$
$B = \mu_o.e / 4.\pi.R_n = \cancel{4.\pi.R_n}.m_e/e^2 . e / \cancel{4.\pi.R_n} = m_e/e = 1/RC$ kg/C
RC_e is the relative atomic charge of an electron {C/kg}
$B = 1/RC = 5.685634367312E{-}12$ kg/C

A6 Corollaries

Corollary 1: Everything in the universe is composed of electrical and magnetic energy

Corollary 2: Magnetic energy is accrued and travels from positive to negative

Corollary 3: Electrical energy is shared and travels from negative to positive

Corollary 4: Atoms comprise collections of proton-electron pairs

Corollary 5: A neutron is a proton-electron pair united under high temperature and holds 4.0935556113127E-14 J of energy

Corollary 6: Atoms exist in solid/liquid state (attraction) due to the magnetic field generated by its proton-electron pairs

Corollary 7: Atoms exist in gaseous state (repulsion) due to the electrical field generated by its proton-electron pairs

Corollary 8: All matter is either viscous (solid/liquid) or gaseous, dependent upon the dominance of its magnetic or electrical fields

Corollary 9: Mass is non-polar magnetic charge

Corollary 10: Gravity is the attractive force between magnetic charges

Corollary 11: Light is electro-magnetic energy

Corollary 12: Every orbital system *must* have a force-centre

Corollary 13: Isaac Newton's gravitational constant may be defined as follows:
$G = a_o.c^2/\rho_u$ {m^3 / s^2.kg per m^3}
Where: ρ_u is the ultimate density (7.12661E+16 kg/m^3)

Corollary 14: Potential energy remains constant irrespective of distance from its source

Corollary 15: Orbital shape is defined *only* by force-centre mass

Corollary 16: Kinetic energy of a satellite in Newton's laws of orbital motion is exclusive of that generated by its angular velocity

Corollary 17a: PE/KE = -2.(1-e)/(1-e²) for all orbits
Corollary 17b: The potential energy between a satellite and its force-centre is always twice the kinetic energy in the satellite for circular orbits

Corollary 18: The internal pressure of any mass can be calculated using Isaac Newton's force-formula thus:
p = G.m₁.m₂ / A², in which 'A' is the spherical area at radius 'r' from its centre (G must be in its modified form: 8.3862834423E-10 m³ / s².kg).

Corollary 19: The centrifugal force on an orbiting body:
@ the perigee of an ellipse; F_c = F / (1+e)
@ the apogee of an ellipse; F_c = F . (1+e)

Corollary 20: The centrifugal velocity (v_c) of a satellite at any point in an elliptical orbit is:
v_c = ζ.v
Where:
ζ = √[(ƒ.Sin(θ/2)ᵃ + p.Cos(θ/2)ᵃ) / (ƒ.cos(θ/2)ᵃ + p.Sin(θ/2)ᵃ)]
a = √[⁴/₃.π]
v = the satellite's elliptical velocity at the same point.

Corollary 21: A satellite's spin is defined by its force-centre's spin and its sub-satellite orbits

Corollary 22a#: Prograde spin is induced in a satellite by the potential energy between its centre and that of its force-centre
Corollary 22b#: Retrograde spin is induced in a satellite by prograde spin energy in its force-centre
Corollary 22c#: Prograde spin is induced in a force-centre by the sum of the perigee kinetic energies plus apogee potential energies of its satellite(s)
Assuming a satellite's orbit is in the prograde direction#:

Corollary 23: The difference between the spin rates in a satellite's core and its mantle is due to the conflicting influences of corollary 22

Corollary 24a: A satellite's internal heat is generated by the friction due to corollary 23
Corollary 24b: Heat can be generated within a satellite orbiting a force-centre with a period different to the force-centre's period of spin
Corollary 24c: Heat is increased within a satellite that has satellites of its own
Corollary 24d: A gas-planet is a satellite with sufficient internal heat to prevent a surface crust forming

Corollary 25: A satellite's mantle plumes are generated by its internal frictional heat.

Corollary 26: A satellite's magnetic field is generated by the differential spin rates in its core and its mantle.

Corollary 27: The angular difference between true and magnetic axes in a satellite with iron core and mantle that rotate at different rates can be calculated thus:

$a = \text{sign}(|\omega/\omega_m|) \cdot \frac{1}{2}.\sqrt{[\,A\sin(|\omega/\omega_m|)\,]}$

Corollary 28: All stars produce hydrogen gas in the form of proton-electron pairs, by converting proton-electron pairs (in its body-matter) to neutrons using the frictional heat generated by planetary spin

Corollary 29: Any satellite may be replaced in any orbit without altering the orbital shape and period (velocities)

Corollary 30: A black-body is a celestial body of any mass or size that is too cold to emit electro-magnetic radiation

Corollary 31: The force-centre at the heart of the Milky Way galaxy (Hades) has a mass of ≈1.8E+41 kg, a diameter of 3.5E+12 m and is spinning at ≈2E-07 °/s according to NASA data on the Milky Way

Corollary 32: The density of the matter beneath the earth's crust is little more than that of water

Corollary 33: The coupling ratio (φ), that of magnetic to electrical potential energy is; 4.407E-40

Corollary 34: There is insufficient pressure at the core of a minimum celestial black-body to alter the density of matter.

Corollary 35: Satellites in circular orbits generate their own kinetic energy.

Corollary 36a: All heat is radiated
Corollary 36b: Conduction is the transfer of electro-magnetic energy between electrons within matter irrespective of its state
Corollary 35c: Convection is the movement of gaseous atoms to balance electrical repulsive forces (between adjacent atoms) with gravitational forces

Corollary 37: E=mc² applies to potential energy in circular orbits

Corollary 38: Mass remains constant irrespective of velocity

Corollary 39: There is no such thing as a photon
Electrons in free-flight do not emit light – light possesses no mass

Corollary 40: There is no such thing as dark matter in the form of sub-atomic particles

Philosophiæ Naturalis Principia Mathematica Revision IV

A7 Hypotheses

Hypothesis 1: There is no such thing as mass.

Hypothesis 2: The earth's magnetic field reverses when a galactic comet passes sufficiently close to tip the earth on its axis.

Hypothesis 3: Electrons gain kinetic energy from electro-magnetic radiation, but they can only lose it via proton-electron pairing or impact.

Hypothesis 4: A lattice structure is mirrored in the atomic nucleic matrix.

Hypothesis 5: Only atoms of identical nucleic construction can generate lattice structures.

Hypothesis 6: An element's lattice structure also applies to its gaseous form and is responsible for partial pressure.

A8 The Heroes

The heroes of this story, to which I offer my gratitude, are listed below

It is not necessary to identify the invaluable contributions made by each of these contributors, they are all widely known and available in almost every scientific publication in circulation today.

Nicolaus Copernicus (Polish) 1473-1543
William Gilbert (English) 1544-1603
Tyco Brahe (Danish) 1546-1601
Galileo Galilei (Italian) 1564-1642
Johannes Kepler (German) 1571-1630
Christiaan Huygens (Dutch) 1629-1695
Isaac Newton (English) 1642-1727
Edmund Halley (English) 1656-1741
Charles-Augustin de Coulomb (French) 1736-1806
Hans Christian Ørsted (Danish) 1777-1851
Michael Faraday (English) 1791-1867
Josef Stefan (Austria) 1815-1863
James Clerk Maxwell (Scottish) 1831-1879
William Crookes (English) 1832-1919
Ludwig Boltzmann (Austria) 1844-1906
Hendrik Lorentz (Dutch) 1853-1928
Jules Henri Poincaré (French) 1854-1912
Johannes Robert Rydberg (Swedish) 1854-1919
Max Karl Ernst Ludwig Planck (German) 1858-1947

The others that were instrumental in the completion of this book are:

My long-suffering wife (Brigitte) sub-editor and critic

My daughter (Eléonore), who initiated this project

Kenneth Pickering friend & editor, who first suggested that I write it

My thanks go out to all the above each of whom have provided a valuable piece of the puzzle without which the final solution would not have been possible, along with my sincere apologies to anybody I have unintentionally omitted.